MicroRNAs and Cardiovascular Disease

By

Zhiguo Wang

Department of Medicine, Montreal Heart Institute,
University of Montreal, Montreal, Quebec, Canada

CONTENTS

CHAPTERS

Section I: miRNAs in Cardiovascular System

Section II: miRNAs and Cardiac Disease

Section III: miRNAs and Vascular Disease

Section IV: miRNAs and Other Aspects of Cardiac Function

Section V: miRNAs as Therapeutic Targets for Cardiovascular Disease

FOREWORD

Cardiovascular disease has been the major cause of death of human beings; fighting this number one killer has been an endless endeavor of physicians and scientists worldwide. Armed with updated knowledge and improved technology is a basic requirement for wining the fight. In the past few years, we have witnessed rapid evolution of a novel scientific field: micro RNA (miRNA) has entered the macro world of heart. Several milestones in this field are identified. In the year 2005, Srivastava's group reported a critical requirement of miRNAs to cardiac development, being the first documentation of miRNA function in mammalian heart. In 2006, Olsen's group demonstrated the causative involvement of miRNAs in cardiac hypertrophy; representing the starting point of miRNA research in adult heart. This was followed by another six original papers on the same topic in 2007. The third milestone is the discovery, by the collaborative effort from Wang's laboratory and my laboratory, of miRNAs as regulators of cardiac excitability and the associated arrhythmogenesis and sudden cardiac death. To date, more than 400 original research papers on the role of miRNAs in cardiovascular physiology and pathology have been published and new discoveries in this field are now published on weekly basis. We have already had a wealth of information about miRNAs in cardiovascular function and disease and the interest in this subject is still rapidly increasing. There is apparently an urgent need to have a systematic summary of these research data beyond single review articles.

The book, *microRNAs and Cardiovascular Disease*, written by Dr. Zhiguo Wang aims to provide such a summary, which is just timely and necessary. In this book, readers can find comprehensive reviews of diverse subjects related to miRNAs and the cardiovascular system, literature-exhausting collection of currently available data, up-to-date knowledge of potential pathophysiological implications, insightful and thoughtful analysis of existing questions and problems, etc. The book contains 18 chapters with the first two chapters on the basics of miRNAs and expression profiles of miRNAs in the cardiovascular system and the rest each focusing on a particular aspect of cardiovascular physiology and pathology and on the relevant technologies. The book, therefore, can be used as a reference book for researchers and professionals or as a textbook for university students.

Baofeng Yang, MD, PhD
Academician, Chinese Academy of Engineering
Professor, Department of Pharmacology
President, Harbin Medical University
China

PREFACE

Cardiovascular disease remains the major cause of morbidity and mortality; according to statistics, heart failure, the syndrome consequential to many diseases of the cardiovascular system, is estimated to have a prevalence of 1–2% and an annual incidence of 5–10 per 1,000 in the developed countries and is the leading cause of hospitalization in the population over 50 years of age. Worse so, there is a clear tendency of increasing prevalence of cardiovascular disease in this planet particularly in the developing nations. This problem casts enormous health concern and costly socioeconomic burden worldwide. Improving our understanding in depth of cardiac pathophysiology and translating our updated knowledge into the clinical practice through the innovation of novel diagnostics and therapeutics has become the most urgent task of scientists and health professionals.

We have entered post-genome era after the human genome project had been completed years ago. Only <2% of all transcribed bases of the entire human genome constitutes the genetic sequence encoding proteins and the rest of 98% accounting for ~70% of all genes carry the sequences for RNAs not encoding a polypeptide chain that was used to be considered for many years "junk DNA" of no physiologic function; proteins were generally assumed the sole biopolymer capable of regulatory function. Intriguingly, the proportion of transcribed non-protein-coding sequences increases with developmental complexity and is a better indicator of phylogenetic level than the number of protein-coding genes of an organism. It is now known that "junk DNA" encodes non-protein-coding RNAs (ncRNAs) that are involved in determining the expression of protein-coding genes by regulating the activity of that 2% of the genome. These ncRNAs include microRNAs (miRNAs), once ignored completely or overlooked as cellular detritus, which were discovered over a decade ago have recently taken many by surprise because of their widespread expression and diverse functions. The discovery of miRNAs challenges the central dogma of molecular biology that has be held since the latter half of the 20th century. This central tenet teaches the flow of genetic information: ribonucleotiac acid (RNA) functions either as infrastructure (a role performed by housekeeping species of RNA) or merely as an intermediate from the deoxynucleotide sequence of genomic DNA through the transcription process, housed within the nucleus, to the amino acid sequence of proteins, which are synthesized in the cytoplasm by the translation process. It has been firmly established that miRNAs play a central role in regulating the expression of protein-coding genes by translational repression at the post-transcriptional level. Since their discovery in 1993, the importance of miRNAs has steadily gained appreciation and now miRNA biology has exploded into a massive swell of interests with enormous range and potential in almost every biological discipline.

microRNAs and Cardiovascular Disease aims to target a wide range of readers from graduate students to post-doctoral fellow and senior researchers involving cardiovascular and miRNA research fields in universities, research institutions, and pharmaceuticals. The contents of the book are also suitable for cardiologists and other health professionals, who are interested in updating their knowledge and utilizing miRNAs as a complementary and an alternative strategy for clinical diagnosis and therapy of human cardiac disease. It provides the most updated information on miRNAs in the cardiovascular system based on the literature available, along with focused highlight, detailed reanalysis, unbiased commentary, and rational speculation. The book is divided into five sections that contain a total of 18 chapters. The book begins with Section I introducing the basics of miRNAs and expression profiles of miRNAs in the cardiovascular system. From Section II to Section IV, each of the 14 chapters is focused on the role of miRNAs in an independent entity of cardiovascular function or disease. Each single chapter contains three subsections: introduction, main body, and reference. In addition, wherever appropriate, illustrations, flow charts or tables for easier and straightforward understanding of the contents are presented.

Zhiguo Wang
University of Montreal
Canada

CHAPTER 1

miRNA Biology

Abstract: This first chapter of the book aims to provide readers the basic information necessary for better understanding the contents of other chapters. The chapter begins with the stories or history of discovery of miRNAs. From this section, readers can get a sense on how some groundbreaking scientific findings could be achieved from small but mysterious things. In the second section, the whole pathway of biogenesis of miRNAs is described in detail, from transcription to maturation. This is followed by the section on action of miRNAs describing how mature miRNAs are hooked up to the protein complex and guide the complex to target genes to execute repression of gene expression at the post-transcriptional level. Finally, some important issues related to miRNA function are introduced in the section entitled "Some conceptual notes about miRNA function."

DISCOVERY OF miRNA

With the recent advance of research into microRNAs (miRNAs), this category of endogenous small non-coding ribonucleic acids (19-25 nts in length) has rapidly emerged as one of the central regulators of expression of an extensive repertoire of genes. MiRNAs are an abundant RNA species constituting >2% of the predicted human genes (>1000 genes), which regulates ~90% of human protein–coding genes. Some miRNAs are expressed at >1000 copies per cell. Thousands of miRNAs have been identified in several organisms including humans, some of which are registered in miRBase Registry (http://microrna.sanger.ac.uk/registry/; the Wellcome Trust Sanger Institute). Computational prediction suggests even a larger number of miRNAs (~25,000 in humans) existing in mammalian genome that are to be identified [Miranda *et al.*, 2006; Cummins *et al.*, 2006].

The high sequence conservation across metazoan species suggests strong evolutionary pressure and participation of miRNAs in essential biological processes such as cell proliferation, differentiation, apoptosis, metabolism, stress, and the forth [Lewis *et al.*, 2003 & 2005; Jackson & Standart, 2007; Nilsen, 2007; Pillai *et al.*, 2007, Alvarez-Garcia & Miska, 2005; Ambros, 2004]. MiRNAs are also critically involved in a variety of pathological processes including human disease, such as developmental malformations, cancer, cardiovascular disease, neuronal disorders, metabolic disturbance, and viral disease. miRNAs have been considered a part of the epigenetic program in organisms. Discovery process of animal miRNAs can be divided into the following stages

Stage I

The first tip-off that non-coding RNAs (ncRNAs) might actually be functional can be traced back to the early 1980s with the discovery of the enzymatic activity of RNA. ncRNAs can be subdivided into house-keeping and regulatory classes. Housekeeping ncRNAs are usually expressed constitutively and are required for the normal function of the cell (such as transfer, ribosomal, small nuclear, and small nucleolar RNA) and regulatory ncRNAs, or riboregulators, are expressed selectively and affect the expression of other genes at the level of transcription or translation. miRNAs belong to the second category of ncRNAs, the regulatory ncRNAs. A loss-offunction mutation of *lin-4*, a heterochronic gene, capable of controlling the timing of development of nematodes, was reported to lead to an increased number of molts and continued synthesis of larval-specific proteins in *Caenorhabditis elegans* [Chalfie *et al.*, 1981]. Several years later, *lin-4* was demonstrated to function as an inhibitor of 2 other heterochronic genes, *lin-14* and *lin-28*, which halt the larva-to-adult switch, thus relieving their inhibition on *lin-29* and permitting continuation of development [Ambros, 1989]. Subsequently, deletions within the 3'untranslated region (UTR) of *lin-14* mRNA were described to result in gain-of-function mutants, with abnormal accumulation of the assumed LIN-14 protein in the later larval stages [Ruvkun *et al.*, 1991]. The authors speculated that the *lin-4* gene product bound to the 3'UTR of *lin-14* negatively regulating it.

Stage II

The second milesone of miRNA discovery is credited to the pioneer work described by Lee *et al* [1993] and Wightman *et al* [1993] in their effort to search for a protein responsible for the disruption of the timing of larval to adult developmental stages due to the lin-4 mutation in the nematode worm *C. elegans* through forward genetic screening. These authors found that *lin-4* does not code for a protein but produces small RNA transcripts, complementary to the 3'UTR of *lin-14* mRNA. And translation of *lin-14* could be inhibited at post-transcriptional level by the interaction of small *lin-4* RNA transcripts with the 3'UTR of *lin-14* mRNA.

Stage III

These above studies however did not arouse major attention in the scientific community until when the let-7 (lethal-7) mutation, which also resulted in disruption of developmental timing in *C. elegans*, was mapped to another small RNA [Reinhart *et al.*, 2000]. From that point, researchers began to realize that the let-7 miRNA sequence, along with its expression during development, was conserved in animals from arthropods to humans [Pasquinelli *et al.*, 2000], indicating that miRNAs represent an ancient mechanism of gene regulation. At the time, these transcripts were called small temporal (st)RNAs and were considered to be restricted in importance.

Stage IV

Thus, lin-4 represents the first founding member of the miRNA family which can downregulate the protein lin-14, and let-7 is the second miRNA mediating translational repression of lin-41. However, the tidal wave of miRNA that hits the field of biology was not yet stirred up until three hallmark papers had been simultaneously published in the journal *Science* which reported the presence of large numbers of small, noncoding RNAs in *Drosophila*, *Caenorhabditis elegans*, and mammalian cells with similar characteristics to *lin-4* and *let-7*: being capable of regulating gene expression at the post-translational level by base pairing at the 3'UTR of their mRNA targets [Lau *et al.*, 2001; Lee & Ambros, 2001; Lagos-Quintana *et al.*, 2001]. These studies in vertebrates suggested that miRNAs, rather than functioning as decisive regulatory on-off switches, more commonly function to modulate or fine-tune cellular phenotypes by repressing expression of proteins that are inappropriate for a particular cell type or by adjusting protein dosage. Thereafter, this new class of small regulatory RNAs gained its big name miRNA and began to garner interest of scientists worldwide.

BIOGENESIS OF miRNA

Genes for miRNAs, like the protein-coding genes, are located in the chromosomes forming an integral part of the complex genome. Yet, the miRNA genes appear more genomically diverse and dynamic. **(1)** A number of miRNAs are transcribed from multiple copies of their genes; **(2)** Around half of all miRNAs are identified in clusters that can be transcribed as polycistronic primary transcripts containing more than one miRNA in one polycistronic transcript; **(3)** Some miRNAs are encoded by their own genes and others are encoded by the sequences as a part of the host protein-coding genes. Based on the genomic arrangement of miRNA genes, miRNAs can be grouped into two classes:

> **(A)** intergenic miRNAs (miRNA-coding genes located in between protein-coding genes),
>
> **(B)** intragenic miRNAs (miRNA-coding genes located within their host protein-coding genes). Further, the intragenic miRNAs can be divided into the following subclasses:
>
> **(Ba)** intronic miRNAs (miRNA-coding genes located within introns of their host protein-coding genes),
>
> **(Bb)** exonic miRNAs (miRNA-coding genes located within exons of host protein-coding genes),
>
> **(Bc)** 3'UTR miRNAs (miRNA-coding genes located within 3'UTR of host protein-coding genes).
>
> **(Bd)** 5'UTR miRNAs (miRNA-coding genes located within 5'UTR of host protein-coding genes).

According to our analysis, for ~800 human miRNAs registered to the online miRNAs data base miRBase (www.mirbase.org/) currently hosted in the Faculty of Life Sciences, University of Manchester, a majority of miRNAs belong to intergenic and intronic miRNAs being comprised of ~40% and ~50% of the total, respectively, and the other three categories are rare with the exonic miRNAs being ~7.5%, 3'UTR miRNAs being 1.5% and 5'UTR miRNAs being 1%.

Clearly, miRNAs either have their own genes or are associated with their host genes; accordingly, miRNAs are generated by two different but **converging pathways**: the canonical pathway and mirtron pathway [Condorelli *et al.*, 2010] (see also Figs. **1** & **2**).

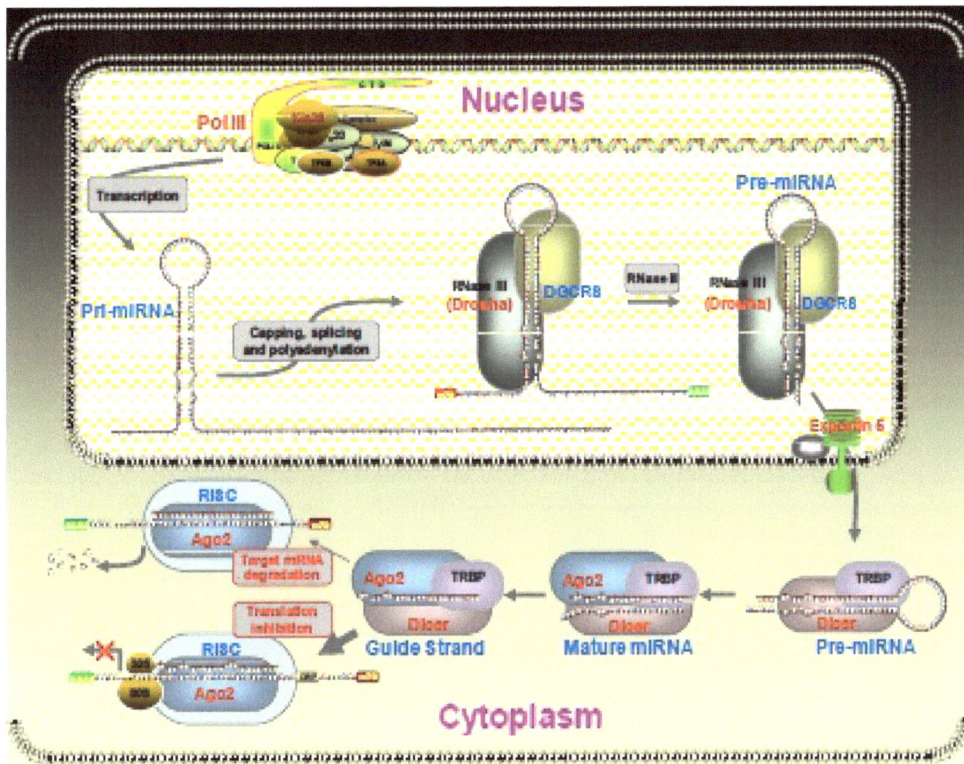

Figure 1: Diagram illustrating the biogenesis pathway of intragenic miRNAs. Intragenic miRNAs are generated by the hairpin structures within host genes (mostly in introns) and they are normally transcribed along with their host genes by Pol II. Pol II: polymerase III, Ago: Argonaute protein-2.

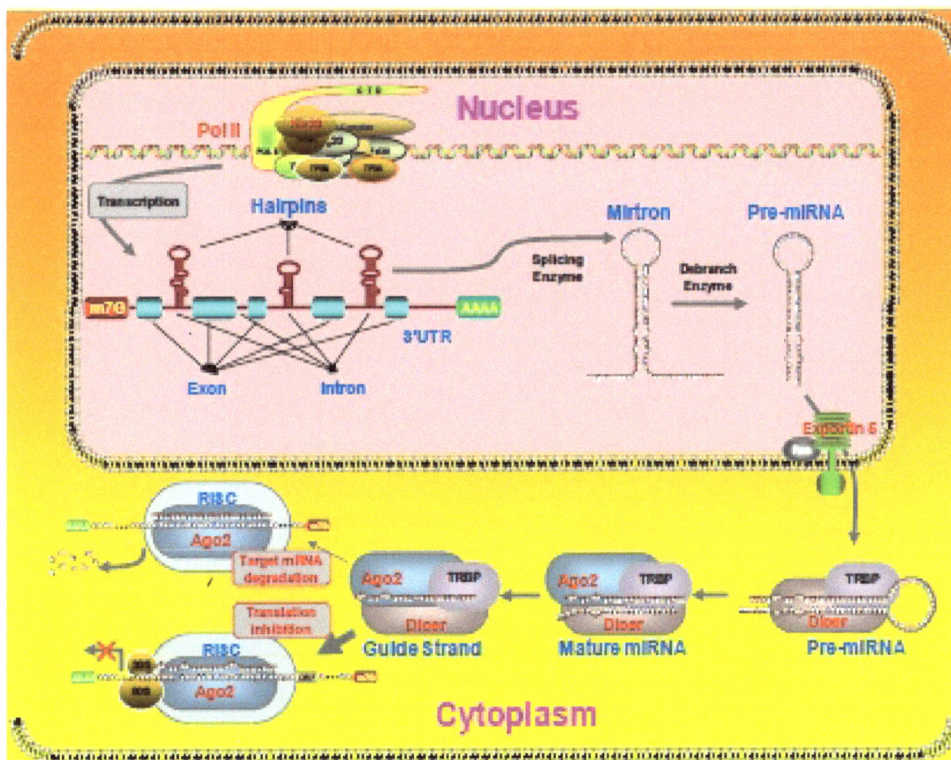

Figure 2: Diagram illustrating the biogenesis pathway of intragenic miRNAs. Intragenic miRNAs are generated by the hairpin structures within host genes (mostly in introns) and they are normally transcribed along with their host genes by Pol II. Pol II: polymerase III, Ago: Argonaute protein-2.

The Canonical Pathway of miRNA Biogenesis

(1) Generation of Primary miRNAs: Transcription of miRNA Genes

The intergenic miRNA genes are first transcribed as long transcripts (up to several thousands of bases long), called primary miRNAs (pri-miRNAs). Transcription of miRNA genes into pri-miRNAs is mostly polymerase II-dependent, though some miRNAs associated with interspersed repetitive DNA elements or derived from viruses are transcribed by another RNA polymerase polymerase III. Like protein-coding genes, the transcription of miRNA genes is tighly controlled by transcription factors and other co-factors [Ying & Lin, 2005]. Intergenic miRNA promoters specifically transcriptional start sites (TSS) have been mapped at distances from 1 kb to as much as 100 kb away from mature miRNA loci [7]. The pri-miRNAs are capped and polyadenylated, and have a characteristic hairpin morphology, comprising a loop and an imperfectly paired stem incorporating the mature miR sequence on one of the strands near the loop [Cullen, 2004; Kim, 2005].

Many of the known miRNAs are expressed as polycistronic transcripts; the genes encoding miRNAs are often found as clustered sequences in the genome or multiple miRNAs on a single pri-miRNA [Tanzer & Stadler, 2004]. This genomic feature confers frequent co-expression of a miRNA with their neighboring miRNAs in the same cluster. For example, the *miR-17~92* cluster, located on the human chromosome 13q31, is a prototypical example of a polycistronic miRNA gene encoding six miRNAs (*miR-17-5p, miR-18, miR-19a, miR-19b, miR-20* and *miR-92*). This cluster has two paralogs, *miR-106a~363* (*miR-106a, miR-18b, miR-19b-2, miR-20b, miR-92a-2* and *miR-363*, located on the X chromosome) and *miR-106b~25* (*miR-106b, miR-93,* and *miR-25*, located on human chromosome 7), which are located on different chromosomes but contain individual miRNAs that are highly similar to those encoded by the *miR-17~92* cluster. The clustered miRNA genes in polycistronic transcripts are likely to be coordinately regulated [Bartel, 2004].

(2) Generation of Precursor miRNAs: Endonuclease Processing of Pri-miRNAs

The pri-miRNAs are processed with the branches flanking the stem-loop structure removed to become precursor miRNAs (pre-miRNAs) by the nuclear RNase endonuclease-III Drosha and its partner DGCR8/Pasha (the DiGeorge syndrome critical region 8, a double-stranded RNA-binding protein) in the nucleus [Lee *et al.*, 2002b; Denli *et al.*, 2004; Gregory *et al.*, 2004; Landthaler *et al.*, 2004]. These pre-miRNAs are ~60 to ~100 nts with a stem-loop or hairpin secondary structure. Specific RNA cleavage by Drosha predetermines the mature miRNA sequence and provides the substrates for subsequent processing steps. Cleavage of a pri-miRNA by microprocessor begins with DGCR8 recognizing the single-stranded RNA (ssRNA)–double-stranded RNA (dsRNA) junction typical of a pri-miRNA [Han *et al.*, 2006]. Then, Drosha is brought close to its substrate through interaction with DGCR8 and cleaves the stem of a pri-miRNA ~11 nt away from the two single-stranded segments. Cleavage by Drosha introduces staggered cuts on each side of the RNA stem, resulting in a 5' phosphate and a two-nucleotide overhang at the 3' end, important for their recognition in later processing.

(3) Nucleus to Cytoplasm Translocation of Pre-miRNAs

Pre-miRNAs then get exported to the cytoplasm from the nucleus through the Ran–GTPdependent nuclear pores exportin-5 [Bohnsack *et al.* 2004; Lund *et al.* 2004; Yi *et al.* 2003]. After a pre-miRNA is exported to the cytoplasm, RanGTP is hydrolyzed by RanGAP to RanGDP, and the pre-miRNA is released from Exp-5.

(4) Generation of Mature miRNAs: Endonuclease Processing of Pre-miRNAs

In the cytoplasm, pre-miRNAs are further processed by the multiprotein RNA–induced silencing complex (miRISC)–loading complex (miRLC). The miRLC is an agglomeration of proteins that removes the loop portion of the pre-miRNA by an enzyme called Dicer associated with transactivation-response element RNA-binding protein (TRBP) and protein activator of the interferon-induced protein kinase (PACT) [Chendrimada *et al.*, 2005; Lee *et al.*, 2006], which is a highly conserved, cytoplasmic RNase III ribonuclease that chops pre-miRNAs into ~22-nt duplexes of mature miRNAs containing a guide strand and a passenger strand (miRNA/miRNA*), with 2-nt overhangs at the 3' termini [Kim, 2005]. Dicer further strips away the passenger strand (miRNA*) from the duplex to leave a mature miRNA and transfers the mature miR from Dicer to another protein of the miRLC, called Argonaute (Ago) [Maniataki & Mourelatos, 2005; Hutvagner *et al.*, 2001 & 2008].

The Mirtron Pathway of miRNA Biogenesis

miRNA precursor-containing introns have recently been designated "mirtrons" [Miranda *et al.*, 2006]. The intronic miRNAs are processed by sharing the same promoter and other regulatory elements of the host genes. They are first transcribed along with their host genes by RNA polymerase II (the canonical pathway) and spliced by spliceosome out of their host genes to form looped intermediates, bypassing the Drosha pathway [Latronico & Condorelli, 2009; Berezikov *et al.*, 2007; Okamura *et al.*, 2007; Ruby *et al.*, 2007]. These intermediates are then debranched and refolded into the typical stem-loop structure with 5′ monophosphates and 3′ 2-nt hydroxyl overhangs, which mimic the structural hallmarks of pre-miRNAs as generated in the canonical pathway. The subsequent steps converge into the canonical miRNA-processing pathway being processed by Dicer [Okamura *et al.*, 2007; Ruby *et al.*, 2007].

The discovery of mirtrons suggests that any RNA, with a size comparable to a pre-miRNA and all the structural features of a pre-miRNA, can be utilized by the miRNA processing machinery, and potentially give rise to a functional miRNA. According to the prediction by Miranda *et al* [2006], the number of mitrons in the human genome may exceed 25,000.

ACTION OF MIRNA

(1) Incorporation of Mature miRNA into miRISC

This is the first step for their function in gene regulation. As already mentioned above, only one strand of miRNA/miRNA*, the guide strand, is successfully incorporated into miRISC, while the other strand, the passenger strand, is eliminated. Strand selection is likely determined by the relative thermodynamic stability of two ends of miRNA duplexes [Khvorova *et al.*, 2003; Schwarz *et al.*, 2003]. The strand with less stability at the 5' end is favorably loaded onto RISC, whereas the passenger strand is released or destroyed. miRISC contains several proteins such as Dicer, TRBP, PACT, and Gemin3, but the components directly associated with miRNAs are Argonaute proteins (Ago). These proteins contain four domains: the N-terminal, PAZ, middle, and Piwi domains. The PAZ domain binds to the 3' end of guide miRNA, while other three domains form a unique structure, creating grooves for target mRNA and guide miRNA interactions [Liu *et al.*, 2004; Song *et al.*, 2004; Ma *et al.*, 2005; Parker *et al.*, 2005]. In mammalian cells, four Ago proteins have been identified, all of which can bind to endogenous miRNAs [Meister & Tuschl, 2004]. Despite the sequence similarity among these Ago proteins, only Ago2 exhibits endonuclease activity to slice complementary mRNA sequences between positions 10 and 11 from the 5' end of guide strand miRNA. Therefore, human Ago2 is a component not only of miRISC but also of siRISC (siRNA-induced silencing complex), a RISC assembled with exogenously introduced siRNA. The roles of various Ago proteins in mammalian RISC are ambiguous, but the division of labor among Ago proteins in *Drosophila* is well defined. *Drosophila* Ago1 and Ago2 have been shown by biochemical and genetic evidence to participate in two separate pathways: Ago1 interacts with miRNA in translational repression, whereas Ago2 associates with siRNA for target cleavage [Carmell *et al.*, 2002; Okamura *et al.*, 2004].

(2) Binding of miRISC to Target Gene

Subsequent binding of a miRNA in the miRISC to sites that seem to be predominantly present on the 3' untranslated region (3'UTR) of its target mRNA through a Watson-Crick basepairing mechanism with its 5'–end 2 to 8 nts exactly complementary to recognition motif within the target. Thus, the actual function of a miRNA is to serve as the target-recognition component of the miRISC. This 5'–end 2 to 8 nt region is termed "seed sequence" or "seed site" for it is critical for miRNA actions [Lewis *et al.*, 2003 & 2005]. Partial complementarity with the rest of the sequence of a miRNA creates bulges and mismatches in the miRNA:mRNA heteroduplex and plays a role in producing post-transcriptional regulation of gene expression, presumably by stabilizing the miRNA:mRNA interaction. Moreover, the mid and 3'–end regions of a miRNA may also be important for forming miRISC. Studies have shown that in addition to 3'UTR, coding region and 5'UTR can also interact with miRNAs to induce gene silencing [Jopling *et al.*, 2005; Luo *et al.*, 2008; Tay *et al.*, 2008].This rather lax requirement for miRNA:mRNA interaction confers the ability of a single miRNA to target multiple (could be hundreds of) genes that can be part of related cellular processes or pathways.

(3) Execution of Post-Transcriptional Gene Regulation by miRISC

Once seating on a target mRNA, the proteins making up the miRISC inhibit the translation of the protein the target mRNA encodes, in mammalian species. A number of animal miRs have been reported to promote gene silencing through mRNA degradation [Carthew & Sontheimer, 2009; Filipowicz *et al.*, 2008]. It has been believed that translational repression is the default mechanism of miRNA-mediated repression of gene expression [Brodersen *et al.*, 2008; Wang, 2009]: whether this occurs at the initiation or a post-initiation step is still being debated. Whether a miRISC inhibits translation or induce degradation of its target mRNA or both depends upon at least the following factors (see Fig. **3**):

(1)   The overall degree of complementarity of the binding site,

(2)   the number of recognition motif corresponding to 5'–end 2 to 8 nts of the miRNA, and

(3)   the accessibility of the bindings sites (as determined by free energy states) [Jackson & Standart, 2007; Nilsen, 2007; Pillai *et al.*, 2007].

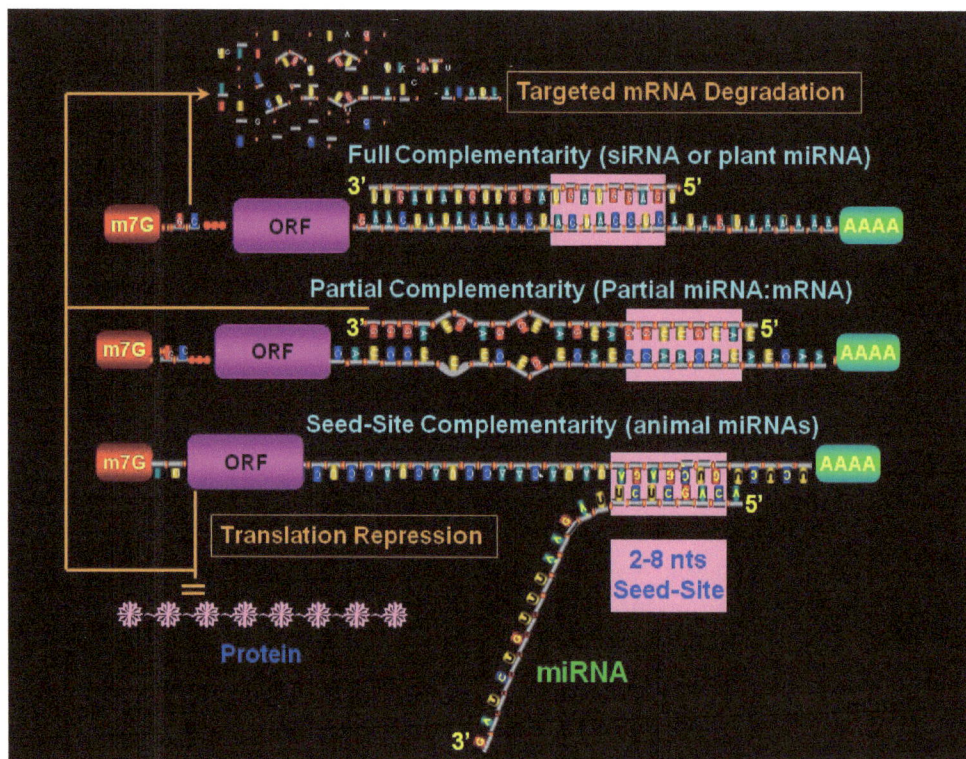

Figure 3: Schematic illustration of mechanisms of action of miRNAs. Full complementarity between a miRNA and its target mRNA (full miRNA:mRNA) results in targeted mRNA cleavage; Seed-site complementarity (Seed-Site miRNA:mRNA) leads to translation inhibition; and partial complementarity (Partial miRNA:mRNA) give rise to both targeted mRNA degradation and protein translation repression.

The greater the degree of complementarity of accessible binding sites, the more likely a miRNA degrades its targeted mRNA. Perfectly complementary targets (**Full miRNA:mRNA interaction**), extending out from the seed site along the first helical turn (11 nt) of the miRNA:mRNA complex, are efficiently silenced by the endonucleolytic cleavage activity of some Argonaute proteins [Hutvágner & Zamore, 2002; Yekta *et al.*, 2004; Davis *et al.*, 2005]. mRNA degradation by miRISC is initiated by deadenylation and decapping of the targeted mRNAs [Pillai *et al.*, 2007].

However, target degradation is a rare phenomenon for miRNAs [Du & Zamore, 2005]; the vast majority of predicted targets in animals are only partially paired (**Partial miRNA:mRNA interaction**) [Lewis *et al.*, 2003 & 2005; Grun *et al.*, 2005; Krek *et al.*, 2005; Rajewsky *et al.*, 2004; Brennecke *et al.*, 2005] and can hardly be cleaved [Haley &

Zamore, 2004]. Mismatched nucleotides in the miRNA:mRNA heteroduplex form bulges that can prevent the cleavage function of Ago2 by moving these nucleotides away from the catalytic site of this endonuclease. Some miRNA has only seed-site complementarity (**Seed-site miRNA:mRNA**). And those miRNAs that display imperfect sequence complementarities with target mRNAs primarily result in translational inhibition [Lewis *et al.*, 2003 & 2005; Jackson & Standart, 2007; Nilsen, 2007; Pillai *et al.*, 2007]. The mechanisms for translational inhibition remain largely unkown, although inhibition of translation initiation has been identified as one such mechanism by several studies [Humphreys *et al.*, 2005; Pillai *et al.*, 2005]. Greater actions may be elicited by a miRNA if it has more than one accessible binding sites in its targeted miRNA, presumably by the cooperative miRNA:mRNA interactions from different sites.

A recent study demonstrated, however, that miRNAs can also act to enhance translation when AU-rich elements and miRNA target sites coexist at proximity in the target mRNA and when the cells are in the state of cell-cycle arrest [Vasudevan *et al.*, 2007]. Moreover, some miRNAs might have motifs that can redirect them to the nucleus [Hwang *et al.*, 2007] where they can mediate transcriptional, rather than the more usual posttranscriptional, gene silencing [Kim *et al.*, 2008].

SOME CONCEPTUAL NOTES ABOUT miRNA FUNCTION

(1) It has been predicted that each single miRNA can have >1000 target genes and each single protein-coding gene can be regulated by multiple miRNAs [Lewis *et al.*, 2003 & 2005; Jackson & Standart, 2007; Nilsen, 2007; Pillai *et al.*, 2007, Alvarez-Garcia & Miska, 2005; Ambros, 2004]. This is at least partially a result of the lax requirement of complementarity for miRNA::mRNA interaction [Lim *et al.*, 2005]. This implies that actions of miRNAs are sequence- or motif-specific, but not gene-specific; different genes can have same binding motifs for a given miRNA and a given gene can have multiple binding motifs for distinct miRNAs. This also confers an advantageous feature of miRNA-based regulation: that is, the ability of single miRNAs to regulate multiple functionally related mRNAs, as shown for the miR-17-5p, which regulates multiple apoptosis-related genes. The targeting of multiple genes that participate in common cellular processes and enables miRNAs to effectively regulate complex intracellular pathways, thereby potentially avoiding redundant mechanisms that might bypass a single inhibited target. *Based on the characteristics of miRNA actions, I postulated that a miRNA should be viewed as a regulator of a cellular function or a cellular program, not of a single gene* [Wang *et al.*, 2008].

(2) Regardless of genomic location and organization, miRNAs function in a distinct yet cooperative manner to regulate cellular processes by coordinately targeting related proteins. miRNAs often belong to families of closely related or identical sequences. We have introduced the "Seed Site" concept proposed by Lewis *et al* [2003 & 2005] in **Chapter 1-3** to define the mechanism of target recognition and action of miRNAs (ie. miRNA:mRNA interactions). Based on this concept, miRNAs possessing a same seed motif (5'-end 2-8 nts) should have a same repertoire of target genes thereby the same cellular function. In the other word, Because of their homology in the seed sequence, the related miRNAs are able to target the same mRNAs, which enhances the efficiency of repression. This concept has indeed been verified by numerous experimental investigations.

For the sake of easiness and clarity in understanding the function of miRNAs, I proposed to categorize miRNAs into families based on their function or seed motifs. According to this classification system, miRNAs with a same seed motif 5'-end 2-8 nts are grouped into the same **miRNA Seed Family**. Further, miRNAs carrying exactly the same seed motif 1-8 nts are grouped into the same **miRNA Seed Subfamily**. For example, miR-17-5p and miR-20b have identical 5'-end 1-8 nts seed sequence CAAAGUGC; miR-520g and miR-520h have ACAAAGUG; miR-20a and miR-106b contains UAAAGUGC; miR-106a, AAAAGUGC; miR-93, miR-372, and miR-520a-e all have AAAGUGCU; miR-519b and miR-519c have AAAGUGCA. Intriguingly, if 7 of 8 nts in the seed motif basepairing with target genes is sufficient to produce post-transcriptional repression (as has already been shown by an enormous volume of studies) [Lewis *et al.*, 2003 & 2005; Pillai *et al.*, 2007], then these six seed motifs should all give the same cellular effects. Based on this view, I consider these miRNAs are all the members of one miRNA seed family while belonging to six different subfamilies.

This classification provides a guideline of pivotal importance for interfering with a cellular process involving gene expression regulation by a multi-member miRNA seed family. Enhancing or inhibiting any one of the members of a miRNA seed family may not be able to elicit, at least not efficient, thorough, changes of gene expression and cellular function. In this case, manipulation of all members of a miRNA seed family is definitely required to achieve a level with sufficient miRNA-promoting or anti-miRNA effects.

We have sorted out all miRNAs registered in miRBase by their seed motifs and are able to categorize these miRNAs into 498 seed families. For convenience, I designate these families according to their 4-7 nts (Table **1**). Some families contain subfamilies with varying number of miRNAs, and some currently contain only one member.

REFERENCES

Abderrahmani A, Plaisance V, Lovis P, Regazzi R. (2006) Mechanisms controlling the expression of the components of the exocytotic apparatus under physiological and pathological conditions. Biochem Soc Trans 34:696–700.

Abelson JF, Kwan KY, O'Roak BJ, Baek DY, Stillman AA, Morgan TM, Mathews CA, Pauls DL, Rasin MR, Gunel M, Davis NR, Ercan-Sencicek AG, Guez DH, Spertus JA, Leckman JF, Dure LS 4th, Kurlan R, Singer HS, Gilbert DL, Farhi A, Louvi A, Lifton RP, Sestan N, State MW. (2005) Sequence variants in SLITRK1 are associated with Tourette's syndrome. Science 310:317–320.

Alvarez-Garcia I, Miska EA. (2005) MicroRNA functions in animal development and human disease. Development 132:4653–4662.

Ambros V, Bartel B, Bartel DP, Burge CB, Carrington JC, Chen X, Dreyfuss G, Eddy SR, Griffiths-Jones S, Marshall M, Matzke M, Ruvkun G, Tuschl T. (2003a) A uniform system for microRNAs annotation. RNA 9:277–279.

Ambros V, Lee RC, Lavanway A, Williams PT, Jewell D. (2003b) MicroRNAs and other tiny endogenous RNAs in *C. elegans*. Curr Biol 13:807−818.

Ambros V. (2004) The functions of animal microRNAs. Nature 431:350–355.

Ambros V. (1989) A hierarchy of regulatory genes controls a larva-to-adult developmental switch in C. elegans. Cell 57:49 –57.

Andersson MG, Haasnoot PC, Xu N, Berenjian S, Berkhout B, Akusjärvi G. (2005) Suppression of RNA interference by adenovirus virus-associated RNA. J Virol 79:9556–9565.

Arisawa T, Tahara T, Shibata T, Nagasaka M, Nakamura M, Kamiya Y, Fujita H, Hasegawa S, Takagi T, Wang FY, Hirata I, Nakano H. (2007) A polymorphism of microRNA 27a genome region is associated with the development of gastric mucosal atrophy in Japanese male subjects. Dig Dis Sci 52:1691–1697.

Bao N, Lye KW, Barton MK. (2004) MicroRNA binding sites in Arabidopsis class III HD-ZIP mRNAs are required for methylation of the template chromosome. Dev Cell 7:653–662.

Barciszewska MZ, Szymański M, Erdmann VA, Barciszewski J. (2000) 5S ribosomal RNA. Biomacromolecules 1:297−302.

Bartel DP. (2004) MicroRNAs: genomics, biogenesis, mechanism, and function. Cell 116:281–297.

Barth S, Pfuhl T, Mamiani A, Ehses C, Roemer K, Kremmer E, Jäker C, Höck J, Meister G, Grässer FA. (2008) Epstein-Barr virus-encoded microRNA miR-BART2 down-regulates the viral DNA polymerase BALF5. Nucleic Acids Res 36:666–675.

Bauersachs J, Thum T. (2007) MicroRNAs in the broken heart. Eur J Clin Invest 37:829–833.

Behlke MA. (2006) Progress towards *in vivo* use of siRNAs. Mol Ther 13:644–670.

Bennasser Y, Le SY, Benkirane M, Jeang KT. (2005) Evidence that HIV-1 encodes an siRNA and a suppressor of RNA silencing. Immunity 22:607–169.

Berezikov E, Chung WJ, Willis J, Cuppen E, Lai EC. (2007) Mammalian mirtron genes. Mol Cell 28:328–336.

Bernstein E, Kim SY, Carmell MA, Murchison EP, Alcorn H, Li MZ, Mills AA, Elledge SJ, Anderson KV, Hannon GJ. (2003) Dicer is essential for mouse development. Nat Genet 35:215–217.

Bertino JR, Banerjee D, Mishra PJ. (2007) Pharmacogenomics of microRNA: a miRSNP towards individualized therapy. Pharmacogenomics 8:1625–1627.

Birmingham A, Anderson EM, Reynolds A, Ilsley-Tyree D, Leake D, Fedorov Y, Baskerville S, Maksimova E, Robinson K, Karpilow J, Marshall WS, Khvorova A. (2006) 3'UTR seed matches, but not overall identity, are associated with RNAi off-targets. Nat Methods 3:199–204.

Boehm M, Slack FJ. (2005) A developmental timing microRNA and its target regulate life span in C. elegans. Science 310:1954–1957.

Boehm M, Slack FJ. (2006) MicroRNA control of lifespan and metabolism. Cell Cycle 5:837–840.

Bohnsack MT, Czaplinski K, Gorlich D. (2004) Exportin 5 is a RanGTP-dependent dsRNA-binding protein that mediates nuclear export of pre-miRNAs. RNA 10:185–191.

Bommer GT, Gerin I, Feng Y, Kaczorowski AJ, Kuick R, Love RE, Zhai Y, Giordano TJ, Qin ZS, Moore BB, MacDougald OA, Cho KR, Fearon ER. (2007) p53-mediated activation of miRNA34 candidate tumor-suppressor genes. Curr Biol 17:1298–1307.

Bottoni A, Piccin D, Tagliati F, Luchin A, Zatelli MC, degli Uberti EC. (2005) miR-15a and miR-16-1 down-regulation in pituitary adenomas. J Cell Physiol 204:280–285.

Brantl S. (2002) Antisense-RNA regulation and RNA interference. Biochimica et Biophysica Acta 1575:15–25.

Brantl S. (2007) Regulatory mechanisms employed by cis-encoded antisense RNAs. Curr Opin Microbiol 10:102-109.

Breaker RR. (2008) Complex riboswitches. Science 319:1795–1797.

Brennecke J, Stark A, Russell RB, Cohen SM (2005) Principles of microRNA-target recognition. PLoS Biol 3:e85.

Brodersen P, Sakvarelidze-Achard L, Bruun-Rasmussen M, Dunoyer P, Yamamoto YY, Sieburth L, Voinnet O. (2008) Widespread translational inhibition by plant miRNAs and siRNAs. Science 320:1185–1190.

Buck AH, Santoyo-Lopez J, Robertson KA, Kumar DS, Reczko M, Ghazal P. (2007) Discrete clusters of virus-encoded micrornas are associated with complementary strands of the genome and the 7.2-kilobase stable intron in murine cytomegalovirus. J Virol 81:13761–13770.

Burnside J, Bernberg E, Anderson A, Lu C, Meyers BC, Green PJ, Jain N, Isaacs G, Morgan RW. (2006) Marek's disease virus encodes MicroRNAs that map to meq and the latency-associated transcript. J Virol 80:8778–8786.

Cai X, Cullen BR. (2006) Transcriptional origin of Kaposi's sarcoma-associated herpesvirus microRNAs. J Virol 80:2234–2242.

Cai X, Schäfer A, Lu S, Bilello JP, Desrosiers RC, Edwards R, Raab-Traub N, Cullen BR. (2006) Epstein-Barr virus microRNAs are evolutionarily conserved and differentially expressed. PLoS Pathog 2:e23.

Calin GA, Croce CM. (2006) MicroRNA-Cancer connection: the beginning of a new tale. Cancer Res 66:7390–7394.

Calin GA, Dumitru CD, Shimizu M, Bichi R, Zupo S, Noch E, Aldler H, Rattan S, Keating M, Rai K, Rassenti L, Kipps T, Negrini M, Bullrich F, Croce CM. (2002) Frequent deletions and down-regulation of microRNA genes miR15 and miR16 at 13q14 in chronic lymphocytic leukemia. Proc Natl Acad Sci USA 99:15524–15529.

Calin GA, Ferracin M, Cimmino A, Di Leva G, Shimizu M, Wojcik SE, Iorio MV, Visone R, Sever NI, Fabbri M, Iuliano R, Palumbo T, Pichiorri F, Roldo C, Garzon R, Sevignani C, Rassenti L, Alder H, Volinia S, Liu CG, Kipps TJ, Negrini M, Croce CM. (2005) A MicroRNA signature associated with prognosis and progression in chronic lymphocytic leukemia. N Engl J Med 353:1793–1801.

Calin GA, Liu CG, Sevignani C, Ferracin M, Felli N, Dumitru CD, Shimizu M, Cimmino A, Zupo S, Dono M, Dell'Aquila ML, Alder H, Rassenti L, Kipps TJ, Bullrich F, Negrini M, Croce CM. (2004a) MicroRNA profiling reveals distinct signatures in B cell chronic lymphocytic leukemias. Proc Natl Acad Sci USA 101:11755–11760.

Calin GA, Sevignani C, Dumitru CD, Hyslop T, Noch E, Yendamuri S, Shimizu M, Rattan S, Bullrich F, Negrini M, Croce CM. (2004b) Human microRNA genes are frequently located at fragile sites and genomic regions involved in cancers. Proc Natl Acad Sci USA 101:2999–3004.

Cantalupo P, Doering A, Sullivan CS, Pal A, Peden KW, Lewis AM, Pipas JM. (2005) Complete nucleotide sequence of polyomavirus SA12. J Virol 79:13094–13104.

Carè A, Catalucci D, Felicetti F, Bonci D, Addario A, Gallo P, Bang ML, Segnalini P, Gu Y, Dalton ND, Elia L, Latronico MV, Høydal M, Autore C, Russo MA, Dorn GW, Ellingsen O, Ruiz-Lozano P, Peterson KL, Croce CM, Peschle C, Condorelli G. (2007) MicroRNA-133 controls cardiac hypertrophy. Nat Med 13:613–618.

Carmell MA, Xuan Z, Zhang MQ, Hannon GJ. (2002) The Argonaute family: Tentacles that reach into RNAi, developmental control, stem cellmaintenance, and tumorigenesis. Genes Dev 16:2733–2742.

Carthew RW, Sontheimer EJ. (2009) Origins and mechanisms of miRNAs and siRNAs. Cell 136:642–655.

Caudy AA, Myers M, Hannon GJ, Hammond SM. (2002) Fragile X-related protein and VIG associate with the RNA interference machinery, Genes Dev 16:2491–2496.

Chalfie M, Horvitz HR, Sulston JE. (1981) Mutations that lead to reiterations in the cell lineages of C. elegans. Cell 24:59–69..

Chan JA, Krichevsky AM, Kosik KS. (2005) MicroRNA-21 is an antiapoptotic factor in human glioblastoma cells. Cancer Res 65:6029–6033.

Chang TC, Wentzel EA, Kent OA, Ramachandran K, Mullendore M, Lee KH, Feldmann G, Yamakuchi M, Ferlito M, Lowenstein CJ, Arking DE, Beer MA, Maitra A, Mendell JT. (2007) Transactivation of miR-34a by p53 broadly influences gene expression and promotes apoptosis. Mol Cell 26:745–752.

Chen JF, Mandel EM, Thomson JM, Wu Q, Callis TE, Hammond SM, Conlon FL, Wang DZ. (2006) The role of microRNA-1 and microRNA-133 in skeletal muscle proliferation and differentiation. Nat Genet 38:228–233.

Chen JF, Murchison EP, Tang R, Callis TE, Tatsuguchi M, Deng Z, Rojas M, Hammond SM, Schneider MD, Selzman CH, Meissner G, Patterson C, Hannon GJ, Wang DZ. (2008a) Targeted deletion of Dicer in the heart leads to dilated cardiomyopathy and heart failure. Proc Natl Acad Sci USA 105:2111–2116.

Chen K, Song F, Calin GA, Wei Q, Hao X, Zhang W. (2008b) Polymorphisms in microRNA targets: a gold mine for molecular epidemiology. Carcinogenesis 29:1306–1311.

Cheng AM, Byrom MW, Shelton J, Ford LP. (2005) Antisense inhibition of human miRNAs and indications for an involvement of miRNA in cell growth and apoptosis. Nucleic Acids Res 33:1290–1297.

Cheng Y, Ji R, Yue J, Yang J, Liu X, Chen H, Dean DB, Zhang C. (2007) MicroRNAs are aberrantly expressed in hypertrophic heart. Do they play a role in cardiac hypertrophy? Am J Pathol 170:1831–1840.

Chendrimada TP, Gregory RI, Kumaraswamy E, Norman J, Cooch N, Nishikura K, Shiekhattar R. (2005) TRBP recruits the Dicer complex to Ago2 for microRNA processing and gene silencing. Nature 436:740–744.

Chung CH, Bernard PS, Perou CM. (2002) Molecular portraits and the family tree of cancer. Nat Genet 32:533–540.

Chuang JC, Jones PA. (2007) Epigenetics and microRNAs. Pediatr Res 61:24R–29R.

Ciafre SA, Galardi S, Mangiola A, Ferracin M, Liu CG, Sabatino G, Negrini M, Maira G, Croce CM, Farace MG. (2005) Extensive modulation of a set of microRNAs in primary glioblastoma. Biochem Biophys Res Commun 334:1351–1358.

Cimmino A, Calin GA, Fabbri M, Iorio MV, Ferracin M, Shimizu M, Wojcik SE, Aqeilan RI, Zupo S, Dono M, Rassenti L, Alder H, Volinia S, Liu CG, Kipps TJ, Negrini M, Croce CM. (2005) miR-15 and miR-16 induce apoptosis by targeting BCL2. Proc Natl Acad Sci USA 102:13944–13949.

Clements-Jewery H, Hearse DJ, Curtis MJ. (2005) Phase 2 ventricular arrhythmias in acute myocardial infarction: a neglected target for therapeutic antiarrhythmic drug development and for safety pharmacology evaluation. Br J Pharmacol 145:551–564.

Clop A, Marcq F, Takeda H, Pirottin D, Tordoir X, Bibé B, Bouix J, Caiment F, Elsen JM, Eychenne F, Larzul C, Laville E, Meish F, Milenkovic D, Tobin J, Charlier C, Georges M. (2006) A mutation creating a potential illegitimate microRNA target site in the myostatin gene affects muscularity in sheep. Nat Genet 38:813–818.

Condorelli G, Latronico MV, Dorn GW 2nd. (2010) microRNAs in heart disease: putative novel therapeutic targets? Eur Heart J 31:649–658.

Corney DC, Flesken-Nikitin A, Godwin AK, Wang W, Nikitin AY. (2007) MicroRNA-34b and MicroRNA-34c are targets of p53 and cooperate in control of cell proliferation and adhesion-independent growth. Cancer Res 67:8433–8438.

Corsten MF, Miranda R, Kasmieh R, Krichevsky AM, Weissleder R, Shah K. (2007) MicroRNA-21 knockdown disrupts glioma growth *in vivo* and displays synergistic cytotoxicity with neural precursor cell delivered S-TRAIL in human gliomas. Cancer Res 67:8994–9000.

Costa FF. (2007) Non-coding RNAs: lost in translation? Gene 386:1–10.

Costa Y, Speed RM, Gautier P, Semple CA, Maratou K, Turner JM, Cooke HJ. (2006) Mouse MAELSTROM: the link between meiotic silencing of unsynapsed chromatin and microRNA pathway? Hum Mol Genet 15:2324–2334.

Costinean S, Zanesi N, Pekarsky Y, Tili E, Volinia S, Heerema N, Croce CM. (2006) Pre-B cell proliferation and lymphoblastic leukemia/high-grade lymphoma in E(mu)-miR155 transgenic mice. Proc Natl Acad Sci USA 103:7024–7029.

Cox DN, Chao A, Baker J, Chang L, Qiao D, Lin H. (1998) A novel class of evolutionarily conserved genes defined by piwi are essential for stem cell self-renewal. Genes Dev 12:3715–3727.

Cuellar TL, McManus MT. (2005) MicroRNAs and endocrine biology. J Endocrinol 187:327–332.

Cui C, Griffiths A, Li G, Silva LM, Kramer MF, Gaasterland T, Wang XJ, Coen DM. (2006a) Prediction and identification of herpes simplex virus 1-encoded microRNAs. J Virol 80:5499–5508.

Cui Q, Yu Z, Purisima E, Wang E. (2006b) Principles of microRNA regulation of a human cellular signaling network. Mol Systems Biol 2:46–52.

Cullen BR. (2004) Transcription and processing of human microRNA precursors. Mol Cell 16:861–865.

Cummins JM, He Y, Leary RJ, Pagliarini R, Diaz LA Jr, Sjoblom T, Barad O, Bentwich Z, Szafranska AE, Labourier E, Raymond CK, Roberts BS, Juhl H, Kinzler KW, Vogelstein B, Velculescu VE. (2006) The colorectal microRNAome. Proc Natl Acad Sci USA 103:3687–3692.

Dalmay T. (2008) MicroRNAs and cancer. J Inter Med 263:366–375.

Darnell DK, Kaur S, Stanislaw S, Konieczka JH, Yatskievych TA, Antin PB. (2006) MicroRNA expression during chick embryo development. Dev Dyn 235:3156–3165.

Davis E, Caiment F, Tordoir X, Cavaillé J, Ferguson-Smith A, Cockett N, Georges M, Charlier C. (2005) RNAi-mediated allelic trans-interaction at the imprinted Rtl1/Peg11 locus. Curr Biol 15:743–749.

Denli AM, Tops BB, Plasterk RH, Ketting RF, Hannon GJ. (2004) Processing of primary microRNAs by the Microprocessor complex. Nature 432:231–235.

Deshpande G, Calhoun G, Schedl P. (2005) Drosophila Argonaute-2 is required early in embryogenesis for the assembly of centric/centromeric heterochromatin, nuclear division, nuclear migration, and germ-cell formation. Genes Dev 19:1680–1685.

Diederichs S, Haber DA. (2006) Sequence variations of microRNAs in human cancer: alterations in predicted secondary structure do not affect processing. Cancer Res 66:6097–6104.

Dölken L, Perot J, Cognat V, Alioua A, John M, Soutschek J, Ruzsics Z, Koszinowski U, Voinnet O, Pfeffer S. (2007) Mouse cytomegalovirus microRNAs dominate the cellular small RNA profile during lytic infection and show features of posttranscriptional regulation. J Virol 81:13771–13782.

Du T, Zamore PD. (2005) microPrimer: the biogenesis and function of microRNA. Development 132:4645–4652.

Duan R, Pak C, Jin P. (2007) Single nucleotide polymorphism associated with mature miR-125a alters the processing of pri-miRNA. Hum Mol Genet 16:1124–1131.

Dunn W, Trang P, Zhong Q, Yang E, van Belle C, Liu F. (2005) Human cytomegalovirus expresses novel microRNAs during productive viral infection. Cell Microbiol 7:1684–1695.

Eis PS, Tam W, Sun L, Chadburn A, Li Z, Gomez MF, Lund E, Dahlberg JE. (2005) Accumulation of miR-155 and BIC RNA in human B cell lymphomas. Proc Natl Acad Sci USA 102:3627–3632.

Esau CC. (2008) Inhibition of microRNA with antisense oligonucleotides. Methods 44:55-60.

Esau C, Davis S, Murray SF, Yu XX, Pandey SK, Pear M, Watts L, Booten SL, Graham M, McKay R, Subramaniam A, Propp S, Lollo BA, Freier S, Bennett CF, Bhanot S, Monia BP. (2006) miR-122 regulation of lipid metabolism revealed by *in vivo* antisense targeting. Cell Metab 3:87–98.

Esau C, Kang X, Peralta E, Hanson E, Marcusson EG, Ravichandran LV, Sun Y, Koo S, Perera RJ, Jain R, Dean NM, Freier SM, Bennett CF, Lollo B, Griffey R. (2004) MicroRNA-143 regulates adipocyte differentiation. J Biol Chem 279:52361–52365.

Esau CC, Monia BP. (2007) Therapeutic potential for microRNAs. Adv Drug Delivery Rev 59:101–114.

Farh KK-H, Grimson A, Jan C, Lewis BP, Johnston WK, Lim LP, Burge CB, Bartel DP. (2005) The widespread impact of mammalian microRNAs on mRNA repression and evolution. Science 310:1817–1821.

Fedorov Y, Anderson EM, Birmingham A, Reynolds A, Karpilow J, Robinson K, Leake D, Marshall WS, Khvorova A. (2006) Off-target effects by siRNA can induce toxic phenotype. RNA 12:1188–1196.

Fernandez-Velasco M, Goren N, Benito G, Blanco-Rivero J, Bosca L, Delgado C. (2003) Regional distribution of hyperpolarization-activated current I_f and hyperpolarization-activated cyclic nucleotide-gated channel mRNA expression in ventricular cells from control and hypertrophied rat hearts. J Physiol 553:395–405.

Filipowicz W, Bhattacharyya SN, Sonenberg N. (2008) Mechanisms of posttranscriptional regulation by microRNAs: are the answers in sight? Nat Rev Genet 9:102–114.

Fjose A, Drivenes O. (2006) RNAi and microRNAs: from animal models to disease therapy. Birth Defects Res C Embryo Today 78:150–171.

Fujita S, Ito T, Mizutani T, Minoguchi S, Yamamichi N, Sakurai K, Iba H. (2008) miR-21 Gene expression triggered by AP-1 is sustained through a double-negative feedback mechanism. J Mol Biol 378:492–504.

Furnari FB, Adams MD, Pagano JS. (1993) Unconventional processing of the 30 termini of the Epstein-Barr virus DNA polymerase mRNA. Proc Natl Acad Sci USA 90:378–382.

Gao H, Xiao J, Sun Q, Lin H, Bai Y, Yang L, Yang B, Wang H, Wang Z. (2006) A single decoy oligodeoxynucleotides targeting multiple oncoproteins produces strong anticancer effects. Mol Pharmacol 70:1621–1629.

Georges M, Coppieters W, Charlier C. (2007) Polymorphic miRNA-mediated gene regulation: contribution to phenotypic variation and disease. Curr Opin Genet Dev 17:166–176.

Gatignol A, Lainé S, Clerzius G. (2005) Dual role of TRBP in HIV replication and RNA interference: viral diversion of a cellular pathway or evasion from antiviral immunity? Retrovirology 2:65.

Ghuran AV, Camm AJ. (2001) Ischemic heart disease presenting as arrhythmias. Br Med Bull 59:193–210.

Gillies JK, Lorimer IA. (2007) Regulation of p27Kip1 by miRNA 221/222 in glioblastoma. Cell Cycle 6:2005–2009.

Gillet R, Felden B. (2001) Emerging views on tmRNA-mediated protein tagging and ribosome rescue. Mol Microbiol 42:879–85.

Giraldez AJ, Cinalli RM, Glasner ME, Enright AJ, Thomson JM, Baskerville S, Hammond SM, Bartel DP, Schier AF. (2005) MicroRNAs regulate brain morphogenesis in zebrafish. Science 308:833–838.

Gironella M, Seux M, Xie MJ, Cano C, Tomasini R, Gommeaux J, Garcia S, Nowak J, Yeung ML, Jeang KT, Chaix A, Fazli L, Motoo Y, Wang Q, Rocchi P, Russo A, Gleave M, Dagorn JC, Iovanna JL, Carrier A, Pébusque MJ, Dusetti NJ. (2007) Tumor protein 53-induced nuclear protein 1 expression is repressed by miR-155, and its restoration inhibits pancreatic tumor development. Proc Natl Acad Sci USA 104:16170–16175.

Gottwein E, Cullen BR. (2008) Viral and cellular microRNAs as determinants of viral pathogenesis and immunity. Cell Host Microbe 3:375–387.

Gottwein E, Mukherjee N, Sachse C, Frenzel C, Majoros WH, Chi JT, Braich R, Manoharan M, Soutschek J, Ohler U, Cullen BR. (2007) A viral microRNA functions as an ortholog of cellular miR-155. Nature 450:1096–1099.

Gregory RI, Yan KP, Amuthan G, Chendrimada T, Doratotaj B, Cooch N, Shiekhattar R. (2004) The Microprocessor complex mediates the genesis of microRNAs. Nature 432:235–240.

Grey F, Antoniewicz A, Allen E, Saugstad J, McShea A, Carrington JC, Nelson J. (2005) Identification and characterization of human cytomegalovirus-encoded microRNAs. J Virol 79:12095–12099.

Gribaldo S, Brochier-Armanet C. (2006) The origin and evolution of Archaea: a state of the art. Philos Trans R Soc Lond B Biol Sci 361:1007–1022.

Grimm D, Streetz KL, Jopling CL, Storm TA, Pandey K, Davis CR, Marion P, Salazar F, Kay MA. (2006) Fatality in mice due to oversaturation of cellular microRNA/short hairpin RNA pathways. Nature 441:537–541.

Grishok A, Pasquinelli AE, Conte D, Li N, Parrish S, Ha I, Baillie DL, Fire A, Ruvkun G, Mello CC. (2001) Genes and mechanisms related to RNA interference regulate expression of the small temporal RNAs that control *C. elegans* developmental timing. Cell 106:23–34.

Grun D, Wang YL, Langenberger D, Gunsalus KC, Rajewsky N. (2005) microRNAs target predictions across seven Drosophila species and comparison to mammalian targets. PLoS Comput Biol 1:e13.

Grundhoff A, Sullivan CS, Ganem D. (2006) A combined computational and microarray-based approach identifies novel microRNAs encoded by human gamma-herpesviruses. RNA 12:733–750.

Gunawardane LS, Saito K, Nishida KM, Miyoshi K, Kawamura Y, Nagami T, Siomi H, Siomi MC. (2007) A slicer-mediated mechanism for repeat-associated siRNA 5' end formation in Drosophila. Science 315:1587–1590.

Gupta A, Gartner JJ, Sethupathy P, Hatzigeorgiou AG, Fraser NW. (2006) Anti-apoptotic function of a microRNA encoded by the HSV-1 latency-associated transcript. Nature 442:82–85.

Haasch D, Chen YW, Reilly RM, Chiou XG, Koterski S, Smith ML, Kroeger P, McWeeny K, Halbert DN, Mollison KW, Djuric SW, Trevillyan JM. (2002) T cell activation induces a noncoding RNA transcript sensitive to inhibition by immunosuppressant drugs and encoded by the proto-oncogene, BIC. Cell Immunol 217:78–86.

Haley B, Zamore PD. (2004) Kinetic analysis of the RNAi enzyme complex. Nat. Struct. Mol. Biol. 11, 599–606.

Hammond SM. (2006) MicroRNAs as oncogenes. Curr Opin Genet Dev 16:4–9.

Han H, Long H, Wang H, Wang J, Zhang Y, Yang B, Wang Z. (2004) Cellular remodeling of apoptosis in response to transient oxidative insult in rat ventricular cell line H9c2: a critical role of the mitochondria death pathway. Am J Physiol 286:H2169–H2182.

Han H, Wang H, Long H, Nattel S, Wang Z. (2001) Oxidative preconditioning and apoptosis in L-cells: Roles of protein kinase B and mitogen-activated protein kinases. J Biol Chem 276:26357–26364.

Han J, Lee Y, Yeom KH, Nam JW, Heo I, Rhee JK, Sohn SY, Cho Y, Zhang BT, Kim VN. (2006) Molecular basis for the recognition of primary microRNAs by the Drosha-DGCR8 complex. Cell 125:887–901.

Harris TA, Yamakuchi M, Ferlito M, Mendell JT, Lowenstein CJ. (2008) MicroRNA-126 regulates endothelial expression of vascular cell adhesion molecule 1. Proc Natl Acad Sci USA 105:1516–1521.

Hatfield SD, Shcherbata HR, Fischer KA, Nakahara K, Carthew RW, Ruohola-Baker H. (2005) Stem cell division is regulated by the microRNA pathway. Nature 435:974–978.

Hayashita Y, Osada H, Tatematsu Y, Yamada H, Yanagisawa K, Tomida S, Yatabe Y, Kawahara K, Sekido Y, Takahashi T. (2005) A polycistronic microRNAs cluster, miR-17-92, is overexpressed in human lung cancers and enhances cell proliferation. Cancer Res 65:9628–9632.

Hayashi K, Chuva de Sousa Lopes SM, Kaneda M, Tang F, Hajkova P, Lao K, O'Carroll D, Das PP, Tarakhovsky A, Miska EA, Surani MA. (2008) MicroRNA biogenesis is required for mouse primordial germ cell development and spermatogenesis. PLoS ONE 3:e1738.

He H, Jazdzewski K, Li W, Liyanarachchi S, Nagy R, Volinia S, Calin GA, Liu CG, Franssila K, Suster S, Kloos RT, Croce CM, de la Chapelle A. (2005a) The role of microRNA genes in papillary thyroid carcinoma. Proc Natl Acad Sci USA 102:19075–19080.

He L, Thomson JM, Hemann MT, Hernando-Monge E, Mu D, Goodson S, Powers S, Cordon-Cardo C, Lowe SW, Hannon GJ, Hammond SM. (2005b) A microRNA polycistron as a potential human oncogene. Nature 435:828–833.

Hino K, Tsuchiya K, Fukao T, Kiga K, Okamoto R, Kanai T, Watanabe M. (2008) Inducible expression of microRNA-194 is regulated by HNF-1alpha during intestinal epithelial cell differentiation. RNA 14:1433–1442.

Hoheisel JD. (2006) Microarray technology: beyond transcript profiling and genotype analysis. Nat Rev Genet 7:200–210.

Horwich MD, Li C Matranga C, Vagin V, Farley G, Wang P, Zamore PD. (2007) The *Drosophila* RNA methyltransferase, DmHen1, modifies germline piRNAs and single-stranded siRNAs in RISC. Current Biology 17:1265–1272.

Hu Z, Chen J, Tian T, Zhou X, Gu H, Xu L, Zeng Y, Miao R, Jin G, Ma H, Chen Y, Shen H. (2008) Genetic variants of miRNA sequences and non-small cell lung cancer survival. J Clin Invest 118:2600–2608.

Humphreys DT, Westman BJ, Martin DI, Preiss T. (2005) MicroRNAs control translation initiation by inhibiting eukaryotic initiation factor 4E/cap and poly(A) tail function. Proc Natl Acad Sci USA 102:16961–16966.

Hutvagner G, McLachlan J, Pasquinelli AE, Ba′lint E, Tuschl T, Zamore PD. (2001) A cellular function for the RNA-interference enzyme Dicer in the maturation of the let-7 small temporal RNA. Science 293:834–838.

Hutvagner G, Simard MJ. (2008) Argonaute proteins: key players in RNA silencing. Nat Rev Mol Cell Biol 9:22–32.

Hutvágner G, Zamore PD. (2002) A microRNA in a multiple-turnover RNAi enzyme complex. Science 297:2056–2060.

Hwang HW, Wentzel EA, Mendell JT. (2007) A hexanucleotide element directs microrNA nuclear import. Science 315:97–100.

Iorio MV, Ferracin M, Liu CG, Veronese A, Spizzo R, Sabbioni S, Magri E, Pedriali M, Fabbri M, Campiglio M, Ménard S, Palazzo JP, Rosenberg A, Musiani P, Volinia S, Nenci I, Calin GA, Querzoli P, Negrini M, Croce CM. (2005) MicroRNA gene expression deregulation in human breast cancer. Cancer Res 65:7065–7070.

Ivey KN, Muth A, Arnold J, King FW, Yeh RF, Fish JE, Hsiao EC, Schwartz RJ, Conklin BR, Bernstein HS, Srivastava D. (2008) MicroRNA regulation of cell lineages in mouse and human embryonic stem cells. Cell Stem Cell 2:219–229.

Izumiya Y, Kim S, Izumi Y, Yoshida K, Yoshiyama M, Matsuzawa A, Ichijo H, Iwao H. (2003) Apoptosis signal-regulating kinase 1 plays a pivotal role in angiotensin II-induced cardiac hypertrophy and remodeling. Circ Res 93:874–883.

Jackson AL, Burchard J, Schelter J, Chau BN, Cleary M, Lim L, Linsley PS. (2006) Widespread siRNA 'off-target' transcript silencing mediated by seed region sequence complementarity. RNA 12:1179–1187.

Jackson AL, Linsley PS. (2004) Noise amidst the silence: off-target effects of siRNAs? Trends Genet 20:521–524.

Jackson AL, Linsley PS. (2003) Expression profiling reveals off-target gene regulation by RNAi. Nat Biotechnol 21:635–637.

Jackson RJ, Standart N. (2007) How do microRNAs regulate gene expression? Sci STKE 23:243–249.

Jay C, Nemunaitis J, Chen P, Fulgham P, Tong AW. (2007) miRNA profiling for diagnosis and prognosis of human cancer. DNA Cell Biol 26:293–300.

Jazdzewski K, Murray EL, Franssila K, Jarzab B, Schoenberg DR, de la Chapelle A. (2008) Common SNP in pre-miR-146a decreases mature miR expression and predisposes to papillary thyroid carcinoma. Proc Natl Acad Sci USA 105:7269–7274.

Ji R, Cheng Y, Yue J, Yang J, Liu X, Chen H, Dean DB, Zhang C. (2007) MicroRNA expression signature and antisense-mediated depletion reveal an essential role of microRNA in vascular neointimal lesion formation. Circ Res 100:1579–1588.

Jin P, Zarnescu DC, Ceman S, Nakamoto M, Mowrey J, Jongens TA, Nelson DL, Moses K, Warren ST. (2004) Biochemical and genetic interaction between the fragile X mental retardation protein and the microRNA pathway. Nat Neurosci 7:113–117.

Johnson SM, Grosshans H, Shingara J, Byrom M, Jarvis R, Cheng A, Labourier E, Reinert KL, Brown D, Slack FJ. (2005) RAS is regulated by the let-7 microRNA family. Cell 120:635–647.

Jopling CL, Yi M, Lancaster AM, Lemon SM, Sarnow P. (2005) Modulation of hepatitis C virus RNA abundance by a liver-specific MicroRNA. Science 309:1577–1581.

Jost N, Virag L, Bitay M, Takacs J, LengyelC, Biliczki P, Nagy Z, Bogats G, Lathrop DA, Papp JG, Varro A. (2005) Restricting excessive cardiac action potential and QT prolongation: A vital role for I_{Ks} in human ventricular muscle. Circulation 112: 1392–1399.

Khvorova A, Reynolds A, Jayasena SD. (2003) Functional siRNAs and miRNAs exhibit strand bias. Cell 115:209–216.

Kim VN (2005) MicroRNA biogenesis: coordinated cropping and dicing. Nat Rev Mol Cell Biol 6:376–385.

Kim J, Krichevsky A, Grad Y, Hayes GD, Kosik KS, Church GM, Ruvkun G. (2004) Identification of many microRNAs that copurify with polyribosomes in mammalian neurons. Proc Natl Acad Sci USA 101:360–365.

Kim DH, Saetrom P, Snøve O Jr, Rossi JJ. (2008) MicrorNA-directed transcriptional gene silencing in mammalian cells. Proc Natl Acad Sci USA 105:16230–16235.

Kirchhoff F, Greenough TC, Brettler DB, Sullivan JL, Desrosiers RC. (1995) Brief report: absence of intact nef sequences in a long-term survivor with nonprogressive HIV-1 infection. N Engl J Med 332:228–232.

Kiss T. (2001) Small nucleolar RNA-guided post-transcriptional modification of cellular RNAs. EMBO J 20:3617–3622.

Kloosterman WP, Plasterk RHA. (2006) The diverse functions of microRNAs in animal development and disease. Dev Cell 11:441–450.

Knight SW, Bass BL. (2001) A role for the RNase III enzyme DCR-1 in RNA interference and germ line development in *Caenorhabditis elegans*. Science 293:2269–2271.

Kosik KS, Krichevsky AM. (2005) The Elegance of the MicroRNAs: A Neuronal Perspective. Neuron 47:779–782.

Krek A, Grün D, Poy MN, Wolf R, Rosenberg L, Epstein EJ, MacMenamin P, da Piedade I, Gunsalus KC, Stoffel M, Rajewsky N. (2005) Combinatorial microRNA target predictions. Nat Genet 37:495–500.

Krichevsky AM, King KS, Donahue CP, Khrapko K, Kosik KS. (2003) A microRNA array reveals extensive regulation of microRNAs during brain development. RNA 9: 1274–1281.

Krützfeldt J, Rajewsky N, Braich R, Rajeev KG, Tuschl T, Manoharan M, Stoffel M. (2005) Silencing of microRNAs *in vivo* with 'antagomirs'. Nature 438:685–689.

Krützfeldt J, Stoffel M (2006) MicroRNAs: a new class of regulatory genes affecting metabolism. Cell Metab 4:9–12.

Kuehbacher A, Urbich C, Zeiher AM, Dimmeler S. (2007) Role of Dicer and Drosha for endothelial microRNA expression and angiogenesis. Circ Res 101:59–68.

Kumar A. (2007) The silent defense: micro-RNA directed defense against HIV-1 replication. Retrovirology 4:26.

Kumar MS, Lu J, Mercer KL, Golub TR, Jacks T. (2007) Impaired microRNA processing enhances cellular transformation and tumorigenesis. Nat Genet 39:673-677.

Kuehbacher A, Urbich C, Zeiher AM, Dimmeler S. (2007) Role of Dicer and Drosha for endothelial microRNA expression and angiogenesis. Circ Res 101:59–68.

Kuwabara T, Hsieh J, Nakashima K, Taira K, Gage FH. (2004) A small modulatory dsRNA specifies the fate of adult neural stem cells. Cell 116:779−793.

Kwon C, Han Z, Olson EN, Srivastava D. (2005) MicroRNA1 influences cardiac differentiation in Drosophila and regulates Notch signaling. Proc Natl Acad Sci USA 102:18986–18991.

Lagos-Quintana M, Rauhut R, Lendeckel W, Tuschl T. (2001) Identification of novel genes coding for small expressed RNAs. Science 294:853–858.

Lagos-Quintana M, Rauhut R, Yalcin A, Meyer J, Lendeckel W, Tuschl T. (2002) Identification of tissue-specific microRNAs from mouse. Curr Biol 12:735–739.

Lakshmipathy U, Love B, Goff LA, Jörnsten R, Graichen R, Hart RP, Chesnut JD. (2007) MicroRNA expression pattern of undifferentiated and differentiated human embryonic stem cells. Stem Cells Dev 16:1003–1016.

Landgraf P, Rusu M, Sheridan R, Sewer A, Iovino N, Aravin A, Pfeffer S, Rice A, Kamphorst AO, Landthaler M, Lin C, Socci ND, Hermida L, Fulci V, Chiaretti S, Foà R, Schliwka J, Fuchs U, Novosel A, Müller RU, Schermer B, Bissels U, Inman J, Phan Q, Chien M, Weir DB, Choksi R, De Vita G, Frezzetti D, Trompeter HI, Hornung V, Teng G, Hartmann G, Palkovits M, Di Lauro R, Wernet P, Macino G, Rogler CE, Nagle JW, Ju J, Papavasiliou FN, Benzing T, Lichter P, Tam W, Brownstein MJ, Bosio A, Borkhardt A, Russo JJ, Sander C, Zavolan M, Tuschl T. (2007) A mammalian microRNA expression atlas based on small RNA library sequencing. Cell 129:1401–1414.

Landi D, Gemignani F, Naccarati A, Pardini B, Vodicka P, Vodickova L, Novotny J, Försti A, Hemminki K, Canzian F, Landi S. (2008) Polymorphisms within micro-RNA-binding sites and risk of sporadic colorectal cancer. Carcinogenesis 29:579–584.

Landthaler M, Yalcin A, Tuschl T. (2004) The human DiGeorge syndrome critical region gene 8 and Its D. melanogaster homolog are required for miRNA biogenesis. Curr Biol 14:2162–2167.

Lau NC, Lim LP, Weinstein E, Bartel DP. (2001) An abundant class of tiny RNAs with probable regulatory roles in *Caenorhabditis elegans*. Science 294:858–862.

Lecellier CH, Dunoyer P, Arar K, Lehmann-Che J, Eyquem S, Himber C, Saïb A, Voinnet O. (2005) A cellular microRNA mediates antiviral defense in human cells. Science 308:557–560.

Lee RC, Ambros V. (2001) An extensive class of small RNAs in *Caenorhabditis elegans*. Science 294:862–864.

Lee RC, Feinbaum RL, Ambros V. (1993) The C. elegans heterochronic gene lin-4 encodes small RNAs with antisense complementarity to lin-14. Cell 75:843–854.

Lee RC, Feinbaum RL, Ambros V. (1993) The *C. elegans* heterochronic gene lin-4 encodes small RNAs with antisense complementarity to lin-14. Cell 75:843–854.

Lee Y, Hur I, Park SY, Kim YK, Suh MR, Kim VN. (2006) The role of PACT in theRNA silencing pathway. EMBO J 25:522–532.

Lee Y, Jeon K, Lee JT, Kim S, Kim VN. (2002a) MicroRNA maturation: stepwise processing and subcellular localization. EMBO J 21:4663–4670.

Lee TI, Rinaldi NJ, Robert F, Odom DT, Bar-Joseph Z, Gerber GK, Hannett NM, Harbison CT, Thompson CM, Simon I, Zeitlinger J, Jennings EG, Murray HL, Gordon DB, Ren B, Wyrick JJ, Tagne JB, Volkert TL, Fraenkel E, Gifford DK, Young RA. (2002b) Transcriptional regulatory networks in Saccharomyces cerevisiae. Science 298:799–804.

Lewis BP, Shih IH, Jones-Rhoades MW, Bartel DP, Burge CB. (2003) Prediction of mammalian microRNA targets. Cell 115:787–798.

Lewis BP, Burge CB, Bartel DP. (2005) Conserved seed pairing, often flanked by adenosines, indicates that thousands of human genes are microRNA targets. Cell 120:15–20.

Li M, Jones-Rhoades MW, Lau NC, Bartel DP, Rougvie AE. (2005) Regulatory mutations of mir-48, a C. elegans let-7 family MicroRNA, cause developmental timing defects. Dev Cell 9:415–422.

Lim LP, Lau NC, Garrett-Engele P, Grimson A, Schelter JM, Castle J, Bartel DP, Linsley PS, Johnson JM. (2005) Microarray analysis shows that some microRNAs downregulate large numbers of target mRNAs. Nature 433:769–773.

Lin SL, Chang SJ, Ying SY. (2006) Transgene-like animal models using intronic microRNAs. Methods Mol Biol 342:321–334.

Lin X, Ruan X, Anderson MG, McDowell JA, Kroeger PE, Fesik SW, Shen Y. (2005) siRNA-mediated off-target gene silencing triggered by a 7 nt complementation. Nucleic Acids Res 33:4527–4535.

Liu DW, Antzelevitch C. (1995) Characteristics of the delayed rectifier current (I_{Kr} and I_{Ks}) in canine ventricular epicardial, midmyocardial, and endocardial myocytes. A weaker I_{Ks} contributes to the longer action potential of the M cell. Circ Res 76: 351–365.

Liu Y, Mochizuki K, Gorovsky MA. (2004) Histone H3 lysine 9 methylation is required for DNA elimination in developing macronuclei in Tetrahymena. Proc Natl Acad Sci USA 101:1679−1684.

Liu N, Williams AH, Kim Y, McAnally J, Bezprozvannaya S, Sutherland LB, Richardson JA, Bassel-Duby R, Olson EN. (2007) An intragenic MEF2-dependent enhancer directs muscle-specific expression of microRNAs 1 and 133. Proc Natl Acad Sci USA 104:20844−20849.

Llave C, Kasschau KD, Rector MA, Carrington JC. (2002) Endogenous and silencing-associated small RNAs in plants. Plant Cell 14:1605–1619.

Lu S, Cullen BR. (2004) Adenovirus VA1 noncoding RNA can inhibit small interfering RNA and MicroRNA biogenesis. J Virol 78:12868–12876.

Lu J, Getz G, Miska EA, Alvarez-Saavedra E, Lamb J, Peck D, Sweet-Cordero A, Ebert BL, Mak RH, Ferrando AA, Downing JR, Jacks T, Horvitz HR, Golub TR. (2005) MicroRNA expression profiles classify human cancers. Nature 435:834–838.

Lu Y, Thomson JM, Wong HY, Hammond SM, Hogan BL. (2007) Transgenic over-expression of the microRNA miR-17-92 cluster promotes proliferation and inhibits differentiation of lung epithelial progenitor cells. Dev Biol 310:442–453.

Lu Y, Zhou J, Xu C, Lin H, Xiao J, Wang Z, Yang B. (2008) AK/STAT and PI3K/AKT pathways form a mutual transactivation loop and afford lasting resistance to oxidative stress-induced apoptosis in cardiomyocytes. Cell Physiol Biochem 21:305–314.

Lund E, Guttinger S, Calado A, Dahlberg JE, Kutay U. (2004) Nuclear export of microRNA precursors. Science 303:95–98.

Luo X, Lin H, Lu Y, Li B, Xiao J, Yang B, Wang Z. (2007) Transcriptional activation by stimulating protein 1 and post-transcriptional repression by muscle-specific microRNAs of I_{Ks}-encoding genes and potential implications in regional heterogeneity of their expressions. J Cell Physiol 212:358–367.

Luo X, Lin H, Pan Z, Xiao J, Zhang Y, Lu Y, Yang B, Wang Z. (2008) Overexpression of Sp1 and downregulation of miR-1/miR-133 activates re-expression of pacemaker channel genes HCN2 and HCN4 in hypertrophic heart. J Biol Chem 283:20045–20052.

Lustig AJ. (1999) Crisis intervention: The role of telomerase. Proc Natl Acad Sci USA 96:3339–3341.

Ma JB, Yuan YR, Meister G, Pei Y, Tuschl T, Patel DJ. (2005) Structural basis for 50-endspecific recognition of guide RNA by the A. fulgidus Piwi protein. Nature 434:666–670.

Mack GS. (2007) MicroRNA gets down to business. Nat Technol 25:631–638.

Maniataki E, Mourelatos Z. (2005) A human, ATP-independent, RISC assembly machine fueled by pre-miRNA. Genes Dev 19:2979–2990.

McKinsey TA, Olson EN. (2005) Toward transcriptional therapies for the failing heart: chemical screens to modulate genes. J Clin Invest 115:538–546.

Mayr C, Hemann MT, Bartel DP. (2007) Disrupting the pairing between let-7 and Hmga2 enhances oncogenic transformation. Science 315:1576–1579.

Meister G, Tuschl T. (2004) Mechanisms of gene silencing by double-stranded RNA. Nature 431:343–349.

Mello CC, Conte Jr D. (2004) Revealing the world of RNA interference. Nature 431:338–342.

Meng F, Henson R, Wehbe-Janek H, Smith H, Ueno Y, Patel T. (2007) The MicroRNA let-7a modulates interleukin-6-dependent STAT-3 survival signaling in malignant human cholangiocytes. J Biol Chem 282:8256–8264.

Mersey BD, Jin P, Danner DJ. (2005) Human microRNA (miR29b) expression controls the amount of branched chain a-ketoacid dehydrogenase complex in a cell, Hum Mol Genet 14:3371–3377.

Michael MZ, O' Connor SM, van Holst Pellekaan NG, Young GP, James RJ. (2003) Reduced accumulation of specific microRNAs in colorectal neoplasia. Mol Cancer Res 1:882–891.

Miranda KC, Huynh T, Tay Y, Ang YS, Tam WL, Thomson AM, Lim B, Rigoutsos I. (2006) A pattern-based method for the identification of MicroRNA binding sites and their corresponding heteroduplexes. Cell 126:1203–1217.

Mishra PJ, Humeniuk R, Mishra PJ, Longo-Sorbello GS, Banerjee D, Bertino JR. (2007) A miR-24 microRNA binding-site polymorphism in dihydrofolate reductase gene leads to methotrexate resistance. Proc Natl Acad Sci USA 104:13513–13518.

Mishra PJ, Mishra PJ, Banerjee D, Bertino JR. (2008) MiRSNPs or MiR-polymorphisms, new players in microRNA mediated regulation of the cell: Introducing microRNA pharmacogenomics. Cell Cycle 7:853–858.

Miska EA, Alvarez-Saavedra E, Townsend M, Yoshii A, Sestan N, Rakic P, Constantine-Paton M, Horvitz HR. (2004) Constantine-Paton, H.R. Horitz, Microarray analysis of microRNA expression in the developing mammalian brain. Genome Biol 5:R68.

Mochizuki K, Gorovsky MA. (2004a) Conjugationspecific small RNAs in Tetrahymena have predicted properties of scan (scn) RNAs involved in genome rearrangement. Genes Dev 18:2068–2073.

Mochizuki K, Gorovsky MA. (2004b) Small RNAs in genome rearrangement in Tetrahymena. Curr Opin Genet Dev 14:181–187.

Mochizuki K, Fine NA, Fujisawa T, Gorovsky MA. (2002) Analysis of a piwi-related gene implicates small RNAs in genome rearrangement in tetrahymena. Cell 110:689–699.

Moffat J, Sabatini DM. (2006) Building mammalian signalling pathways with RNAi screens. Nat Rev Mol Cell Biol 7:177–187.

Morita S, Horii T, Kimura M, Goto Y, Ochiya T, Hatada I. (2007) One Argonaute family member, Eif2c2 (Ago2), is essential for development and appears not to be involved in DNA methylation. Genomics 89:687–696.

Motsch N, Pfuhl T, Mrazek J, Barth S, Grässer FA. (2007) Epstein-Barr virus-encoded latent membrane protein 1 (LMP1) induces the expression of the cellular microRNA miR-146a. RNA Biol 4:131–137.

Mourelatos Z, Dostie J, Paushkin S, Sharma A, Charroux B, Abel L, Rappsilber J, Mann M, Dreyfuss G. (2002) miRNPs: a novel class of Ribonucleo proteins containing numerous microRNAs. Genes Dev 16:720–728.

Mudhasani R, Zhu Z, Hutvagner G, Eischen CM, Lyle S, Hall LL, Lawrence JB, Imbalzano AN, Jones SN. (2008) Loss of miRNA biogenesis induces p19Arf-p53 signaling and senescence in primary cells. J Cell Biol 181:1055–1063.

Murchison EP, Stein P, Xuan Z, Pan H, Zhang MQ, Schultz RM, Hannon GJ. (2007) Critical roles for Dicer in the female germline. Genes Dev 21:682–693.

Nair V, Zavolan M. (2006) Virus-encoded microRNAs: novel regulators of gene expression. Trends Microbiol 14:169–715.

Nelson PT, Baldwin DA, Kloosterman WP, Kauppinen S, Plasterk RH, Mourelatos Z. (2006) RAKE and LNA-ISH reveal microRNA expression and localization in archival human brain. RNA 12:187–191.

Nilsen TW. (2007) Mechanisms of microRNA-mediated gene regulation in animal cells. Trends Genet 23:243–249.

Niu Z, Li A, Zhang SX, Schwartz RJ. (2007) Serum response factor micromanaging cardiogenesis. Curr Opin Cell Biol 19:618–627.

O'Carroll D, Mecklenbrauker I, Das PP, Santana A, Koenig U, Enright AJ, Miska EA, Tarakhovsky A. (2007) A Slicer-independent role for Argonaute 2 in hematopoiesis and the microRNA pathway. Genes Dev 21:1999–2004.

O'Connell RM, Taganov KD, Boldin MP, Cheng G, Baltimore D. (2007) MicroRNA-155 is induced during the macrophage inflammatory response. Proc Natl Acad Sci USA 104:1604–1609.

Okamura K, Hagen JW, Duan H, Tyler DM, Lai EC. (2007) The mirtron pathway generates microRNA-class regulatory RNAs in Drosophila. Cell 130:89–100.

Okamura K, Ishizuka A, Siomi H, Siomi MC. (2004) Distinct roles for Argonaute proteins in small RNA-directed RNA cleavage pathways. Genes Dev 18:1655–1666.

Omoto S, Fujii YR. (2005) Regulation of human immunodeficiency virus 1 transcription by nef microRNA. J Gen Virol 86:751–755.

Omoto S, Ito M, Tsutsumi Y, Ichikawa Y, Okuyama H, Brisibe EA, Saksena NK, Fujii YR. (2004) HIV-1 nef suppression by virally encoded microRNA. Retrovirology 1:44.

Otsuka M, Jing Q, Georgel P, New L, Chen J, Mols J, Kang YJ, Jiang Z, Du X, Cook R, Das SC, Pattnaik AK, Beutler B, Han J. (2007) Hypersusceptibility to vesicular stomatitis virus infection in Dicer1-deficient mice is due to impaired miR24 and miR93 expression. Immunity 27:123–134.

Otsuka M, Zheng M, Hayashi M, Lee JD, Yoshino O, Lin S, Han J. (2008) Impaired microRNA processing causes corpus luteum insufficiency and infertility in mice. J Clin Invest 118:1944–1954.

Pannucci JA, Haas ES, Hall TA, Harris JK, Brown JW. (1999) RNase P RNAs from some Archaea are catalytically active. Proc Natl Acad Sci USA 96:7803–7808.

Park MY, Wu G, Gonzalez-Sulser A, Vaucheret H, Poethig RS. (2005) Nuclear processing and export of microRNAs in Arabidopsis. Proc Natl Acad Sci USA 102:3691–3696.

Parker JS, Roe SM, Barford D. (2005) Structural insights into mRNA recognition from a PIWI domain-siRNA guide complex. Nature 434:663–666.

Pasquinelli AE, Reinhart BJ, Slack F, Martindale MQ, Kuroda MI, Maller B, Hayward DC, Ball EE, Degnan B, Muller P, Spring J, Srinivasan A, Fishman M, Finnerty J, Corbo J, Levine M, Leahy P, Davidson E, Ruvkun G. (2000) Conservation of the sequence and temporal expression of let-7 heterochronic regulatory RNA. Nature 408:86–89.

Pedersen IM, Cheng G, Wieland S, Volinia S, Croce CM, Chisari FV, David M. (2007) Interferon modulation of cellular microRNAs as an antiviral mechanism. Nature 449:919–922.

Peragine A, Yoshikawa M, Wu G, Albrecht HL, Poethig RS. (2004) SGS3 and SGS2/SDE1/RDR6 are required for juvenile development and the production of transacting siRNAs in Arabidopsis. Genes Dev 18:2368–2379.

Perreault J, Perreault J-P, Boire G. (2007) Ro-associated Y RNAs in metazoans: evolution and diversification. Mol Biol Evol 24:1678–1689.

Pfeffer S, Sewer A, Lagos-Quintana M, Sheridan R, Sander C, Grässer FA, van Dyk LF, Ho CK, Shuman S, Chien M, Russo JJ, Ju J, Randall G, Lindenbach BD, Rice CM, Simon V, Ho DD, Zavolan M, Tuschl T. (2005) Identification of microRNAs of the herpesvirus family. Nat Methods 2:269–276.

Pfeffer S, Zavolan M, Grässer FA, Chien M, Russo JJ, Ju J, John B, Enright AJ, Marks D, Sander C, Tuschl T. (2004) Identification of virus-encoded microRNAs. Science 304:734–736.

Pillai RS, Bhattacharyya SN, Artus CG, Zoller T, Cougot N, Basyuk E, Bertrand E, Filipowicz W. (2005) Inhibition of translational initiation by Let-7 MicroRNA in human cells. Science 309:1573–1576.

Pillai RS, Bhattacharyya SN, Filipowicz W. (2007) Repression of protein synthesis by miRNAs: how many mechanisms? Trends Cell Biol 17:18–126.

Plaisance V, Abderrahmani A, Perret-Menoud V, Jacquemin P, Lemaigre F, Regazzi R. (2006) MicroRNA-9 controls the expression of Granuphilin/Slp4 and the secretory response of insulin-producing cells. J Biol Chem 281:26932–26942.

Poy MN, Spranger M, Stoffel M. (2007) microRNAs and the regulation of glucose and lipid metabolism. Diabetes Obes Metab Suppl 2:67–73.

Rajewsky N, Socci ND. (2004) Computational identification of microRNA targets. Dev Biol 267:529–535.

Randall G, Panis M, Cooper JD, Tellinghuisen TL, Sukhodolets KE, Pfeffer S, Landthaler M, Landgraf P, Kan S, Lindenbach BD, Chien M, Weir DB, Russo JJ, Ju J, Brownstein MJ, Sheridan R, Sander C, Zavolan M, Tuschl T, Rice CM. (2007) Cellular cofactors affecting hepatitis C virus infection and replication. Proc Natl Acad Sci USA 104:12884–12889.

Raveche ES, Salerno E, Scaglione BJ, Manohar V, Abbasi F, Lin YC, Fredrickson T, Landgraf P, Ramachandra S, Huppi K, Toro JR, Zenger VE, Metcalf RA, Marti GE. (2007) Abnormal microRNA-16 locus with synteny to human 13q14 linked to CLL in NZB mice. Blood 109:5079–5086.

Raver-Shapira N, Marciano E, Meiri E, Spector Y, Rosenfeld N, Moskovits N, Bentwich Z, Oren M. (2007) Transcriptional activation of miR-34a contributes to p53-mediated apoptosis. Mol Cell 26:731–743.

Reinhart BJ, Slack FJ, Basson M, Pasquinelli AE, Bettinger JC, Rougvie AE, Horvitz HR, Ruvkun G. (2000) The 21-nucleotide let-7 RNA regulates developmental timing in Caenorhabditis elegans. Nature 403:901–906.

Ricke DO, Wang S, Cai R, Cohen D. (2006) Genomic approaches to drug discovery. Curr Opin Chem Biol 10:303–308.

Rodríguez M, Lucchesi BR, Schaper J. (2002) Apoptosis in myocardial infarction. Ann Med 34:470–479.

Rodriguez A, Vigorito E, Clare S, Warren MV, Couttet P, Soond DR, van Dongen S, Grocock RJ, Das PP, Miska EA, Vetrie D, Okkenhaug K, Enright AJ, Dougan G, Turner M, Bradley A. (2007) Requirement of bic/microRNA-155 for normal immune function. Science 316:608–611.

Rosenkranz-Weiss P, Tomek RJ, Mathew J, Eghbali M. (1994) Gender-specific differences in expression of mRNAs for functional and structural proteins in rat ventricular myocardium. J Mol Cell Cardiol 26:261–270.

Ruby JG, Jan C, Bartel DP. (2007) Intronic microRNA precursors that bypass Drosha processing. Nature 448:83–86.

Ruvkun G. (2001) Molecular biology. Glimpses of a tiny RNA world. Science 294:797–799.

Ruvkun G, Wightman B, Burglin T, Arasu P. (1991) Dominant gain-of-function mutations that lead to misregulation of the C. elegans heterochronic gene lin-14, and the evolutionary implications of dominant mutations in pattern-formation genes. Dev Suppl 1:47–54.

Saetrom P, Snøve O Jr, Rossi JJ. (2007) Epigenetics and microRNAs. Pediatr Res 61:17R–23R.

Saito Y, Liang G, Egger G, Friedman JM, Chuang JC, Coetzee GA, Jones PA. (2006) Specific activation of microRNA-127 with downregulation of the proto-oncogene BCL6 by chromatin-modifying drugs in human cancer cells. Cancer Cell 9:435–443.

Sall A, Liu Z, Zhang HM, Yuan J, Lim T, Su Y, Yang D. (2008) MicroRNAs-based therapeutic strategy for virally induced diseases. Curr Drug Discov Technol 5:49–58.

Salvi R, Garbuglia AR, Di Caro A, Pulciani S, Montella F, Benedetto A. (1998) Grossly defective nef gene sequences in a human immunodeficiency virus type 1-seropositive long-term nonprogressor. J Virol 72:3646–3657.

Samols MA, Hu J, Skalsky RL, Renne R. (2005) Cloning and identification of a microRNA cluster within the latency-associated region of Kaposi's sarcoma-associated herpesvirus. J Virol 79:9301–9305.

Saunders MA, Liang H, Li WH. (2007) Human polymorphism at microRNAs and microRNA target sites. Proc Natl Acad Sci USA 104:3300–3305.

Sayed D, Hong C, Chen IY, Lypowy J, Abdellatif M. (2007) MicroRNAs play an essential role in the development of cardiac hypertrophy. Circ Res 100:416–424.

Scaria V, Hariharan M, Pillai B, Maiti S, Brahmachari SK. (2007) Host-virus genome interactions: macro roles for microRNAs. Cell Microbiol 9:2784–2794.

Scaria V, Jadhav V. (2007) microRNAs in viral oncogenesis. Retrovirology 4:82.

Schat KA, Calnek BW. (1978) Protection against Marek's disease-derived tumor transplants by the nononcogenic SB-1 strain of Marek's disease virus. Infect Immun 22:225–232.

Schratt GM, Tuebing F, Nigh EA, Kane CG, Sabatini ME, Kiebler M, Greenberg ME. (2006) A brain-specific microRNA regulates dendritic spine development. Nature 439:283–289.

Schwarz DS, Hutvagner G, Du T, Xu Z, Aronin N, Zamore PD. (2003) Asymmetry in the assembly of the RNAi enzyme complex. Cell 115:199–208.

Scott GK, Mattie MD, Berger CE, Benz SC, Benz CC. (2006) Rapid alteration of microRNA levels by histone deacetylase inhibition. Cancer Res 66:1277–1281.

Sempere LF, Freemantle S, Pitha-Rowe I, Moss E, Dmitrovsky E, Ambros V. (2004) Expression profiling of mammalian microRNAs uncovers a subset of brain-expressed microRNAs with possible roles in murine and human neuronal differentiation, Genome Biol 5:R13.

Sethupathy P, Borel C, Gagnebin M, Grant GR, Deutsch S, Elton TS, Hatzigeorgiou AG, Antonarakis SE. (2007) Human microRNA-155 on chromosome 21 differentially interacts with its polymorphic target in the AGTR1 3' untranslated region: a mechanism for functional single-nucleotide polymorphisms related to phenotypes. Am J Hum Genet 81:405–413.

Shen J, Ambrosone CB, DiCioccio RA, Odunsi K, Lele SB, Zhao H. (2008) A functional polymorphism in the miR-146a gene and age of familial breast/ovarian cancer diagnosis. Carcinogenesis 29:1963–1966.

Shyu AB, Wilkinson MF, van Hoof A. (2008) Messenger RNA regulation: to translate or to degrade. EMBO J 27:471–481.

Si ML, Zhu S, Wu H, Lu Z, Wu F, Mo YY. (2007) miR-21-mediated tumor growth. Oncogene 26:2799–1803.

Silva JM, Hammond SM, Hannon GJ. (2002) RNA interference: a promising approach to antiviral therapy? Trends Mol Med 8:505–508.

Singh SK. (2007) miRNAs: from neurogeneration to neurodegeneration. Pharmacogenomics 8:971–978.

Skalsky RL, Samols MA, Plaisance KB, Boss IW, Riva A, Lopez MC, Baker HV, Renne R. (2007) Kaposi's sarcoma-associated herpesvirus encodes an ortholog of miR-155. J Virol 81:12836–12845.

Slack FJ, Weidhaas JB. (2006) MicroRNAs as a potential magic bullet in cancer. Future Oncol 2:73–82.

Smirnova L, Gräfe A, Seiler A, Schumacher S, Nitsch R, Wulczyn FG. (2005) Regulation of miRNA expression during neural cell specification. Eur J Neurosci 21:1469–1477.

Soifer HS, Rossi JJ, Saetrom P. (2007) MicroRNAs in disease and potential therapeutic applications. Mol Ther 15:2070–2079.

Song JJ, Smith SK, Hannon GJ, Joshua-Tor L. (2004) Crystal structure of Argonaute and its implications for RISC slicer activity. Science 305:1434–1437.

Srivastava D, Thomas T, Lin Q, Kirby ML, Brown D, Olson EN. (1997) Regulation of cardiac mesodermal and neural crest development by the bHLH transcription factor, dHAND. Nat Genet 16:154–160.

Stenvang J, Kauppinen S. (2008) MicroRNAs as targets for antisense-based therapeutics. Expert Opin Biol Ther 8:59–81.

Stilli D, Sgoifo A, Macchi E, Zaniboni M, De Iasio S, Cerbai E, Mugelli A, Lagrasta C, Olivetti G, Musso E. (2001) Myocardial remodeling and arrhythmogenesis in moderate cardiac hypertrophy in rats. Am J Physiol 280:H142–H150.

Suárez Y, Fernández-Hernando C, Pober JS, Sessa WC. (2007) Dicer dependent microRNAs regulate gene expression and functions in human endothelial cells. Circ Res 100:1164–1173.

Sullivan CS, Ganem D. (2005) MicroRNAs and viral infection. Mol Cell 20:3-7.

Sullivan CS, Grundhoff AT, Tevethia S, Pipas JM, Ganem D. (2005) SV40-encoded microRNAs regulate viral gene expression and reduce susceptibility to cytotoxic T cells. Nature 435:682–686.

Szentadrassy N, Banyasz T, Biro T, Szabo G, Toth BI, Magyar J, Lazar J, Varro A, Kovacs L, Nanasi PP. (2005) Apico-basal inhomogeneity in distribution of ion channels in canine and human ventricular myocardium. Cardiovasc Res 65:851–860.

Taganov KD, Boldin MP, Chang KJ, Baltimore D. (2006) NF-kappaB-dependent induction of microRNA miR-146, an inhibitor targeted to signaling proteins of innate immune responses. Proc Natl Acad Sci USA 103:12481–12486.

Takamizawa J, Konishi H, Yanagisawa K, Tomida S, Osada H, Endoh H, Harano T, Yatabe Y, Nagino M, Nimura Y, Mitsudomi T, Takahashi T. (2004) Reduced expression of the let-7 microRNAs in human lung cancers in association with shortened postoperative survival. Cancer Res 64:3753–3756.

Tam W. (2001) Identification and characterization of human BIC, a gene on chromosome 21 that encodes a noncoding RNA. Gene 274:157–167.

Tam W, Ben-Yehuda D, Hayward WS. (1997) Bic, a novel gene activated by proviral insertions in avian leukosis virus-induced lymphomas, is likely to function through its noncoding RNA. Mol Cell Biol 17:1490–1502.

Tanzer A, Stadler PF. (2004) Molecular evolution of a microRNA cluster. J Mol Biol 339:327–335.

Tarasov V, Jung P, Verdoodt B, Lodygin D, Epanchintsev A, Menssen A, Meister G, Hermeking H. (2007) Differential regulation of microRNAs by p53 revealed by massively parallel sequencing: miR-34a is a p53 target that induces apoptosis and G1-arrest. Cell Cycle 6:1586–1593.

Tatsuguchi M, Seok HY, Callis TE, Thomson JM, Chen JF, Newman M, Rojas M, Hammond SM, Wang DZ. (2007) Expression of microRNAs is dynamically regulated during cardiomyocyte hypertrophy. J Mol Cell Cardiol 42:1137–1141.

Tay Y, Zhang J, Thomson AM, Lim B, Rigoutsos I. (2008) MicroRNAs to Nanog, Oct4 and Sox2 coding regions modulate embryonic stem cell differentiation. Nature 455:1124–1128.

Tazawa H, Tsuchiya N, Izumiya M, Nakagama H. (2007) Tumor-suppressive miR-34a induces senescence-like growth arrest through modulation of the E2F pathway in human colon cancer cells. Proc Natl Acad Sci USA 104:15472–15477.

Tea BS, Dam TV, Moreau P, Hamet P, deBlois D. (1999) Apoptosis during regression of cardiac hypertrophy in spontaneously hypertensive rats. Temporal regulation and spatial heterogeneity. Hypertension 34:229–235.

Teiger E, Than VD, Richard L, Wisnewsky C, Tea BS, Gaboury L, Tremblay J, Schwartz K, Hamet P. (1996) Apoptosis in pressure overload-induced heart hypertrophy in the rat. J Clin Invest 97:2891–2897.

Teleman AA, Maitra S, Cohen SM. (2005) Drosophila lacking microRNA miR-278 are defective in energy homeostasis. Genes Dev 20:417–422.

Thai TH, Calado DP, Casola S, Ansel KM, Xiao C, Xue Y, Murphy A, Frendewey D, Valenzuela D, Kutok JL, Schmidt-Supprian M, Rajewsky N, Yancopoulos G, Rao A, Rajewsky K. (2007) Regulation of the germinal center response by microRNA-155. Science 316:604–608.

Thore S, Mayer C, Sauter C, Weeks S, Suck D. (2003) Crystal Structures of the Pyrococcus abyssi Sm Core and Its Complex with RNA. J Biol Chem 278:1239–1247.

Thum T, Catalucci D, Bauersachs J. (2008) MicroRNAs: novel regulators in cardiac development and disease. Cardiovasc Res 79:562–570.

Thum T, Galuppo P, Wolf C, Fiedler J, Kneitz S, van Laake LW, Doevendans PA, Mummery CL, Borlak J, Haverich A, Gross C, Engelhardt S, Ertl G, Bauersachs J. (2007) MicroRNAs in the human heart: a clue to fetal gene reprogramming in heart failure. Circulation 116:258–267.

Triboulet R, Mari B, Lin YL, Chable-Bessia C, Bennasser Y, Lebrigand K, Cardinaud B, Maurin T, Barbry P, Baillat V, Reynes J, Corbeau P, Jeang KT, Benkirane M. (2007) Suppression of microRNA-silencing pathway by HIV-1 during virus replication. Science 315:1579–1582.

Tuddenham L, Wheeler G, Ntounia-Fousara S, Waters J, Hajihosseini MK, Clark I, Dalmay T. (2006) The cartilage specific microRNA-140 targets histone deacetylase 4 in mouse cells. FEBS Lett 580:4214–4217.

van den Berg A, Kroesen BJ, Kooistra K, de Jong D, Briggs J, Blokzijl T, Jacobs S, Kluiver J, Diepstra A, Maggio E, Poppema S. (2003) High expression of B-cell receptor inducible gene BIC in all subtypes of Hodgkin lymphoma. Genes Chromosomes Cancer 37:20–28.

van Rooij E, Sutherland LB, Liu N, Williams AH, McAnally J, Gerard RD, Richardson JA, Olson EN. (2006) A signature pattern of stress-responsive microRNAs that can evoke cardiac hypertrophy and heart failure. Proc Natl Acad Sci USA 103: 18255–18260.

van Rooij E, Sutherland LB, Qi X, Richardson JA, Hill J, Olson EN. (2007) Control of stress-dependent cardiac growth and gene expression by a microRNA. Science 316:575–579.

Vasudevan S, Tong Y, Steitz JA. (2007) Switching from repression to activation: microRNAs can up-regulate translation. Science 318:1931–1934.

Vazquez F, Vaucheret H. (2004) Endogenous trans-acting siRNAs regulate the accumulation of Arabidopsis mRNAs. Mol Cell 16:1–13.

Verduyn SC, Vos MA, van der Zande J, van der Hulst FF, Wellens HJ. (1997) Role of interventricular dispersion of repolarization in acquired torsade-de-pointes arrhythmias: Reversal by magnesium. Cardiovasc Res 34:453–463.

Vidal L, Blagden S, Attard G, de Bono J. (2005) Making sense of antisense. Eur J Cancer 41:2812–2818.

Vigorito E, Perks KL, Abreu-Goodger C, Bunting S, Xiang Z, Kohlhaas S, Das PP, Miska EA, Rodriguez A, Bradley A, Smith KG, Rada C, Enright AJ, Toellner KM, Maclennan IC, Turner M. (2007) microRNA-155 regulates the generation of immunoglobulin class-switched plasma cells. Immunity 27:847–859.

Volinia S, Calin GA, Liu CG, Ambs S, Cimmino A, Petrocca F, Visone R, Iorio M, Roldo C, Ferracin M, Prueitt RL, Yanaihara N, Lanza G, Scarpa A, Vecchione A, Negrini M, Harris CC, Croce CM. (2006) A microRNA expression signature of human solid tumors defines cancer gene targets. Proc Natl Acad Sci USA 103:2257–2261.

Wang Z, Feng J, Shi H, Pond A, Nerbonne JM, Nattel S. (1999) The potential molecular basis of different physiological properties of transient outward K^+ current in rabbit and human hearts. Circ Res 84:551–561.

Wang Z, Luo X, Lu Y, Yang B. (2008) miRNAs at the heart of the matter. J Mol Med 86:772–783.

Wang Z, Yue L, White M, Pelletier G, Nattel S. (1998) Differential expression of inward rectifier potassium channel mRNA in human atrium versus ventricle and in normal versus failing hearts. Circulation 98:2422–2428.

Welch C, Chen Y, Stallings RL. (2007) MicroRNA-34a functions as a potential tumor suppressor by inducing apoptosis in neuroblastoma cells. Oncogene 26:5017–5022.

Wienholds E, Koudijs MJ, van Eeden FJ, Cuppen E, Plasterk RH. (2003) The microRNA-producing enzyme Dicer1 is essential for zebrafish development. Nat Genet 35:217–218.

Wightman B, Ha I, Ruvkun G. (1993) Posttranscriptional regulation of the heterochronic gene lin-14 by lin-4 mediates temporal pattern formation in *C. elegans*. Cell 75:855–562.

Wilfred BR, Wang WX, Nelson PT. (2007) Energizing miRNA research: a review of the role of miRNAs in lipid metabolism, with a prediction that miR-103/107 regulates human metabolic pathways. Mol Genet Metab 91:209–217.

Wurdinger T, Costa FF. (2007) Molecular therapy in the microRNA era. Pharmacogenomics J 7:297–304.

Woodhams MD, Stadler PF, Penny D, Collins LJ. (2007) RNase MRP and the RNA processing cascade in the eukaryotic ancestor. BMC Evolutionary Biology 7:S13.

Xiao J, Luo X, Lin H, Xu C, Gao H, Wang H, Yang B, Wang Z. (2007) MicroRNA miR-133 represses HERG K^+ channel expression contributing to QT prolongation in diabetic hearts. J Biol Chem 282:12363–12367.

Xu C, Lu Y, Lin H, Xiao J, Wang H, Luo X, Li B, Yang B, Wang Z. (2007) The muscle-specific microRNAs miR-1 and miR-133 produce opposing effects on apoptosis via targeting HSP60/HSP70 and caspase-9 in cardiomyocytes. J Cell Sci 120:3045–3052.

Xu P, Vernooy SY, Guo M, Hay BA. (2003) The Drosophila microRNA Mir-14 suppresses cell death and is required for normal fat metabolism. Curr Biol 13:790–795.

Xu T, Zhu Y, Wei QK, Yuan Y, Zhou F, Ge YY, Yang JR, Su H, Zhuang SM. (2008) A functional polymorphism in the miR-146a gene is associated with the risk for hepatocellular carcinoma. Carcinogenesis 29:2126–2131.

Yanaihara N, Caplen N, Bowman E, Seike M, Kumamoto K, Yi M, Stephens RM, Okamoto A, Yokota J, Tanaka T, Calin GA, Liu CG, Croce CM, Harris CC. (2006) Unique microRNA molecular profiles in lung cancer diagnosis and prognosis. Cancer Cell 9:189–198.

Yang B, Lin H, Xiao J, Lu Y, Luo X, Li B, Zhang Y, Xu C, Bai Y, Wang H, Chen G, Wang Z. (2007) The muscle-specific microRNA miR-1 causes cardiac arrhythmias by targeting GJA1 and KCNJ2 genes. Nat Med 13:486–491.

Yang B, Lu Y, Wang Z. (2008) Control of cardiac excitability by microRNAs. Cardiovasc Res 79:571–580.

Yang M, Mattes J. (2008) Discovery, biology and therapeutic potential of RNA interference, microRNA and antagomirs. Pharmacol Ther 117:94–104.

Yekta S, Shih IH, Bartel DP. (2004) MicroRNA-directed cleavage of HOXB8 mRNA. Science 304:594–596.

Yi R, Qin Y, Macara IG, Cullen BR. (2003) Exportin-5 mediates the nuclear export of pre-microRNAs and short hairpin RNAs. Genes Dev 17:3011–3016.

Yin Q, McBride J, Fewell C, Lacey M, Wang X, Lin Z, Cameron J, Flemington EK. (2008) MicroRNA-155 is an Epstein-Barr virus-induced gene that modulates Epstein-Barr virus-regulated gene expression pathways. J Virol 82:5295–5306.

Ying SY, Lin SL. (2005) Intronic microRNAs. Biochem Biophys Res Commun 326:515–520.

Yu B, Yang Z, Li J, Minakhina S, Yang M, Padgett RW, Steward R, Chen X. (2005) Methylation as a crucial step in plant microRNAs biogenesis. Science 307:932–935.

Yu Z, Li Z, Jolicoeur N, Zhang L, Fortin Y, Wang E, Wu M, Shen SH. (2007) Aberrant allele frequencies of the SNPs located in microRNA target sites are potentially associated with human cancers. Nucleic Acids Res 35:4535–4541.

Yu XY, Song YH, Geng YJ, Lin QX, Shan ZX, Lin SG, Li Y. (2008) Glucose induces apoptosis of cardiomyocytes via microRNA-1 and IGF-1. Biochem Biophys Res Commun 376:548–552.

Zhang C. (2008) MicroRNomics: a newly emerging approach for disease biology. Physiol Genomics 33:139–147.

Zhang Y, Xiao J, Wang H, Luo X, Wang J, Villeneuve LR, Zhang H, Bai Y, Yang B, Wang Z. (2006) Restoring depressed HERG K$^+$ channel function as a mechanism for insulin treatment of the abnormal QT prolongation and the associated arrhythmias in diabetic rabbits. Am J Physiol 291:1446–1455.

Zhao Y, Ransom JF, Li A, Vedantham V, von Drehle M, Muth AN, Tsuchihashi T, McManus MT, Schwartz RJ, Srivastava D. (2007) Dysregulation of cardiogenesis, cardiac conduction, and cell cycle in mice lacking miRNA-1-2. Cell 129:303–317.

Zhao Y, Samal E, Srivastava D. (2005) Serum response factor regulates a muscle specific microRNA that targets Hand2 during cardiogenesis. Nature 436:214–220.

Zhu S, Si ML, Wu H, Mo YY. (2007) MicroRNA-21 targets the tumor suppressor gene tropomyosin 1 (TPM1). J Biol Chem 282:14328–14336.

Zidar N, Jera J, Maja J, Dusan S. (2007) Caspases in myocardial infarction. Adv Clin Chem 44:1–33.

CHAPTER 2

Expression Profiles of miRNAs in Heart and Vessel

Abstract: To date, more than 5000 miRNAs have been identified in animals across the lowest and the highest species, some 800 of them are found in humans. To have better understanding of how miRNAs function in the heart, it is necessary to have an idea about which miRNAs exist in the heart. This chapter focuses on cardiovascular-expressed miRNAs or the miRNA profiles in heart and vessel. Specifically, the miRNAs existing in the heart will be introduced by categories: cardiac-specific, muscle-specific, cardiac-enriched, and other cardiac-expressed, in detail on their genomic locations and relative expression levels in normal cardiac tissues. Alterations of miRNA signature under various diseased states of the heart are succinctly described and the precautions in interpreting the profiling data are stated.

INTRODUCTION

Expression of miRNAs is tissue-restricted and the restriction may be in a qualitative form (some miRNAs are expressed exclusively in certain tissue or cell types but not in others) and in a quantitative form (some miRNAs are abundantly expressed only in certain tissue or cell types and modestly in others). While individual miRNAs may not express in a tissue/cell-specific manner, the expression profile of miRNAs appears to be tissue/cell-specific. This implies that a group of miRNAs express only in a certain tissue/cell type or different groups of miRNAs expressed in different tissues/cells. For example, the miRNA expression profile in artery is different from that in heart. The most abundant miRNAs in cardiac muscles are *miR-1, let-7, miR-133, miR-126–3p, miR-30c*, and *miR-26a* [Lagos-Quintana *et al*., 2002]. However, in artery smooth muscles the most abundant miRNAs are *miR-145, let-7, miR-125b, miR-125a, miR-23*, and *miR-143* [Ji *et al*., 2007], though *miR-1* and *miR-133* are also expressed in artery smooth muscles. The differential tissue distributions of miRNAs suggest tissue– or even cell type–specific function of these molecules.

The expression profile of miRNAs is also disease status–dependent. A particular pathological process is associated with the expression of a particular group of miRNAs, a signature expression pattern of miRNAs. These signature patterns could aid in the diagnosis and prognosis of human disease. This notion has been supported by studies on cardiovascular disease and human cancer.

Among ~820 human miRNAs registered in the miRBase, 220 can be detected in heart by real-time RT-PCR [Luo *et al*., 2010]. Among these cardiac-expressed miRNAs, some are cardiac-specific, some are muscle-specific, some are cardiac-enriched, and others are low-abundant in heart.

CARDIAC-EXPRESSED miRNAs

Cardiac-specific miRNAs

In the adult heart, the MYH6 gene, encoding a fast myosin the alpha-myosin heavy chain (αMHC), coexpresses the intronic miR-208a, which regulates the expression of genes encoding two slow myosins (βMHC and Myh7b) and their intronic miRNAs, MYH7/miR-208b and MYH7b/miR-499, respectively [van Rooij *et al*., 2007 & 2009]. The pre-miR-208a is located within intron 28 of the αMHC gene at chromosome 14, the pre-miR-208b is located within intron 29 of the βMHC gene at chromosome 14, and the pre-miR-499 is located within intron 20 of the MYH7B gene at chromosome 20 [Fig. **1**]. Thus, they all belong to mirtons [Miranda *et al*., 2006]. Activation of MYH7b by miR-208a is constitutive, whereas activation of β-MHC also requires stress signals or absence of thyroid hormone [Morkin, 2000; Weiss & Leinwand, 1996]. Activation of the slow myofiber gene program also creates a positive feedback loop via the expression of miR-208b and miR-499, which further reinforce slow muscle gene program [van Rooij *et al*., 2007 & 2009].

In adult human heart, miR-208a is predominantly expressed in ventricular cells whereas miR-208b in atrial cells. It has been noticed that the abundance of mRNAs for αMHC and βMHC is higher in the heart of adult female rats than in that of age-matched male rats [Rosenkranz-Weiss *et al*., 1994]. However, the overall abundance of these three miRNAs in cardiac tissues is low, around the detection margin [Liang *et al*., 2007; Luo *et al*., 2010]. A role for miR-208a is to regulate the shift from αMHC to βMHC during cardiac stress [van Rooij *et al.*, 2007]. miR-208b and miR-499 play redundant roles in the specification of muscle fiber identity by activating slowand repressing fast myofiber gene programs. The actions of these miRNAs are mediated in part by a collection of transcriptional repressors of slow myofiber genes. These findings reveal that myosin genes not only encode the major contractile proteins of muscle, but act more broadly to influence muscle function by encoding a network of intronic miRNAs that control muscle gene expression and performance [van Rooij *et al.*, 2007 & 2009].

Muscle-specific miRNAs

miR-1 and miR-133 are muscle-specific miRNAs preferentially expressed in cardiac and skeletal muscle and have been shown to regulate differentiation and proliferation of these cells. Thery represent the most abundant miRNAs expressed in heart [Liang *et al.*, 2007; Lagos-Quintana *et al.*, 2002; Luo *et al.*, 2010] and among the most highly conserved miRNAs across the species from nematodes and flies to all vertebrates [Latronico *et al.*, 2007].

Although miR-1 and miR-133 form part of the same bicistronic unit, they are expressed as separate transcripts. The miR-1 family is comprised of the miR-1 subfamily and miR-206, the latter of which is not expressed in the heart. The miR-1 subfamily consists of two closely related transcripts, miR-1-1 and miR-1-2, encoded by distinct genes found on chromosomes 20 and 18 in humans, respectively [Fig. 1]. miR-1-1 is an intronic miRNA whereas miR-1-2 is an intergenic miRNA. In flies, transcription of miR-1 is activated in a broad pan-mesodermal domain before gastrulation, whereas the two mouse miR-1 genes are first detected later, at the beginning of muscle differentiation, and then become progressively more expressed. miR-1-1 is first expressed in the inner curvature of the heart loop and in atria, during mammalian development, but becomes ubiquitously expressed in the heart as development continues; on the other hand, miR-1-2 is prevalent in the ventricles [Zhao *et al.*, 2005].

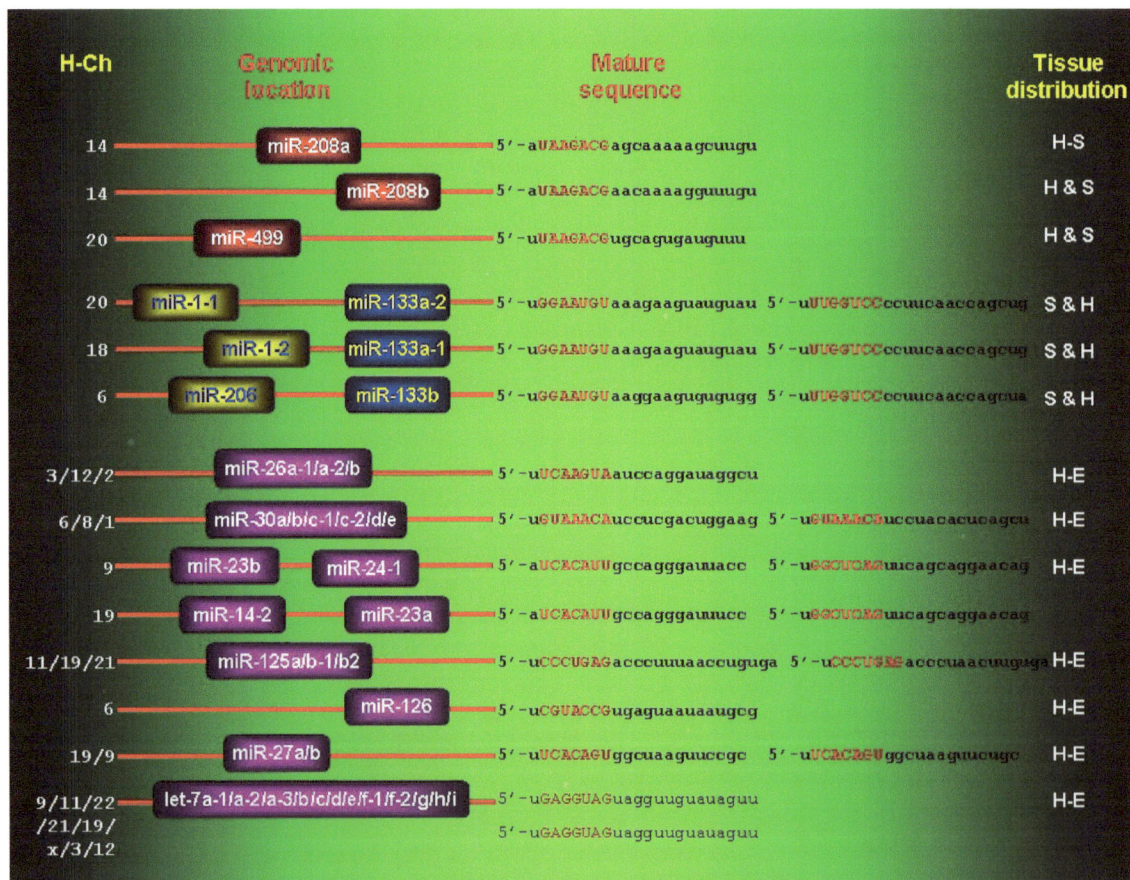

Figure 1: Top 10 most abundant miRNAs expressed in the heart with relative chromosome localization and genomic loci. H-Ch: location on human chromosome; H-S: cardiac-specific; H & S: cardiac-specific with expression in skeletal muscle; S & H: muscle-specific with expression in skeletal and cardiac muscles; H-E: cardiac-enriched.

The miR-133 family is comprised of miR-133a-1, miR-133a-2, and miR-133b. The human miR-133a-1 and miR-133a-2 are expressed in bicistronic units with miR-1-2 and miR-1-1, respectively; thus, miR-133a-1 on chromosome 18 is an intergenic miRNA whereas miR-13a-2 on chromosome 20 is an intronic miRNA. miR-133b is in a bicistronic unit with miR-206. But intriguingly, while miR-206 is not expressed in heart, miR-133b is the top-abundant miRNA in cardiac tissue [Liang *et al.*, 2007]. An ancient genomic duplication is thought to have resulted in 2 distinct loci for the miR-1/miR-133 cluster in vertebrates, with identical mature sequences derived from the duplicated loci [Chen *et al.*, 2006; Rao *et al.*, 2006]. The resulting mature products are either identical or have only 1 base of difference. miR-133 is expressed in heart with increasing levels in developing mouse hearts from day embryonic day (E)12.5 through to at least E18.5 [Carè *et al.*, 2007; Latronico *et al.*, 2007]. Similar to miR-1, muscle-specific expression of miR-133 is regulated by SRF.

Cardiac-enriched miRNAs

Expression of miRNAs in mammalian species under normal conditions is genetically programmed with certain spatial (depending on cell-, tissue-, or organ-type) and temporal (depending on developmental stage) patterns. This property generates the so-called expression signature of a particular tissue. We therefore conducted miRNA microarray analysis of all 718 human miRNAs for their expression in left ventricular tissues of five healthy human individuals. We found 222 out of 718 human miRNAs being expressed in the cardiac tissue [Luo *et al.*, 2010].

According to the results reported by Liang et al [2007] for human heart, the top 20 abundant miRNAs in human heart are miR-1, miR-133a/b, miR-16, miR-100, miR-125a/b, miR-126, miR-145, miR-195, miR-199*, miR-20a/b, miR-21, miR-26a/b, miR-24, miR-23, miR-29a/b, miR-27a/b, miR-30a/b/c, miR-92a/b, miR-99, and let-7a/c/f/g. We verified the expression abundance of several selected miRNAs (miR-1, miR-133a/b, miR-125a/b, miR-30a/b/c, miR-26a/b, miR-24, miR-27a/b, miR-23, miR-29a/b, miR-101, miR-21, miR-150 and miR-328) using RNA samples isolated from left ventricular tissues of healthy human subjects (Fig. **2**) [Luo et al., 2010]. A recent study by Rao et al [2006] reported a similar array of abundant miRNAs in mouse heart. But differences between the two species exist: e.g. miR-1 constitutes ~40% of total miRNA content in mouse, but in human, it is ranked the 2nd most abundant miRNA around 1/3 of the miR-133 level; miR-208 was found to be one of the top 20 abundant miRNA in mouse, but in human miR-208 is rarely expressed; and miR-22, miR-143, miR-499 and miR-451 were considered the most abundant miRNAs in mouse heart but not in human heart.

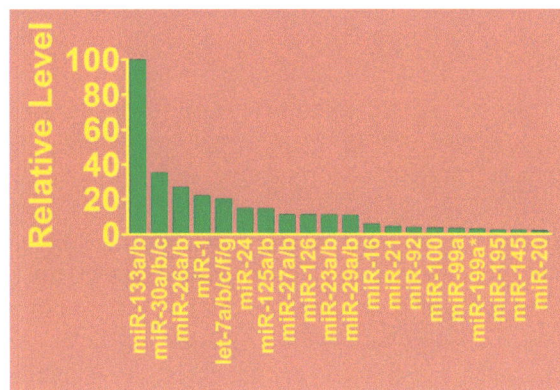

Figure 2: Relative levels of the top 20 most abundantly expressed miRNAs in myocardium, determined by real-time RT-PCR.

We considered the miRNAs with the same seed sequence as one single miRNA for these miRNAs expectedly have the same set of target genes. This consideration might change the relative abundance of miRNAs. For instance, *miR-1* was found more abundant than each of the *miR-30* or *miR-26* isoforms; but was considered less abundant than these latter two miRNAs when the seed family was taken as one miRNA, ranked top 4 after *miR-30a/b/c* (top 2) and *miR-26a/b* (top 3) (Fig. **2**) [Luo *et al.*, 2010].

Noticeably, many of these top 20 abundant cardiac-expressed miRNAs have been studied for their roles in cardiac disease and the relevant targeting mechanisms. Introduction to these results will be given in the following chapters.

Other cardiac-expressed miRNAs

In addition to the miRNAs introduced above, some other miRNAs that are expressed at relatively low levels in myocardium have also been shown to be importantly involved in regulating cardiovascular function. These include miR-195 in cardiac hypertrophy, miR-143 in smooth muscle differentiation, miR-320 in myocardial infarction, miR-9 in cardiac hypertrophy, miR-138 in cardiac development, miR-223 in glucose metabolism, miR-155 in neurohormonal activation, and miR-590 in cardiac fibrosis.

VASCULAR-EXPRESSED miRNAs

Artery Smooth Muscle

Ji *et al* [2007] reported that in rat artery smooth muscles the most abundant miRNAs are *miR-145*, *let-7*, *miR-125b*, *miR-125a*, *miR-23*, and *miR-143*, though *miR-1* and *miR-133* are also expressed in artery smooth muscles.

miRome IN CARDIOVASCULAR DISEASE

A wide spectrum of miRNA expression detection methods has been recently developed for specific purposes [Wang & Yang, 2010]. Among these, miRNA microarray, particularly the LNA-modified array analysis service provide by Exiqon, has been exceptionally useful for miRNA expression profiling or miRNA transcriptome or miRome studies.

The first report of miRome in human heart disease was provided by Thum *et al* [2007] who compared mRNA transcriptome and miRome from four non-failing and six failing hearts. The Ambion miRNA microarrays contained probe sets for 384 miRs. The authors noted significant upregulation of 67 miRNAs, with downregulation of 43 miRNAs, in the failing versus control hearts. The miRome in failing heart was similar to that of fetal heart, suggesting that both mRNA transcriptome and miRome expression in heart failure partially recapitulates that of the embryonic heart. However, the two human studies in which comprehensive mRNA and miRNA signatures were obtained in the same clinical cardiac samples failed to reveal clear reciprocal relationships between upregulated miRNAs and downregulation of their putative mRNA targets [Thum *et al.*, 2007; Naga Prasad *et al.*, 2009]. Nevertheless, miRNAs have important pathophysiological effects on cardiomyocytes, as shown by *in vitro* and *in vivo* manipulation [van Rooij *et al.*, 2006; Carè *et al.*, 2007].

In the review article by Condorelli *et al* [2010], the authors identified a few notes of caution before broad generalizations can be derived from the expression profiling data. First, the data need to be interpreted in the context of evolving platforms for assaying miRNA levels, and the rapid expansion of recognized miRNAs. Older studies assayed a limited number of miRs compared with more recent studies. Second, a cardiac-expressed miRNA is not necessarily a cardiomyocyte-expressed miRNA. Interventions, such as pressure overloading and myocardial infarction that affect the entire heart including interstitial cells, the coronary macro- and micro-vasculature and resident or migratory inflammatory cells, will not produce the same RNA expression signatures as genetic manipulations that directly affect only cardiomyocytes. Finally, genetic models are artificial and even experimental pressure overloading and myocardial infarction are acute interventions that do not precisely mimic the chronically progressive conditions seen in human diseases with the same names. Thus, the miRNA expression profiling data from mouse models identify a large number of miRNAs that are expressed in the heart at relatively high levels, and a subset of these that are subject to dynamic regulation under conditions of myocardial stress or injury. If associations between these experimentally regulated miRNAs and their human counterparts were also observed in clinical heart disease, the implication would be that cross-species conservation of regulation would more strongly suggest biological relevance.

It is anticipated that the cardiovascular miRome will be eventually applied as a biomarker for diagnosis/prognosis of cardiovascular disease; however, it is expectedly affected by many factors: pathology (diseased versus healthy hearts), etiology (different causes of a disease), progression (different stages of a disease), gender, age, etc. For example, the profile of the miRome can be characteristic for a given pathology. A number of reports have described miRNA expression signatures of the heart in the human [Thum *et al.*, 2007; Ikeda *et al.*, 2007; Sucharov *et al.*, 2008; Divakaran & Mann, 2009] and mouse [Lagos-Quintana., 2002; Cheng *et al.*, 2007; van Rooij *et al.*, 2006] and have documented arrays of miRNAs that are misexpressed, positively and/or negatively, with cardiac disease. Modification of the cardiac miRome, however, is not identical for all pathologies of the heart. Etiologic factors are responsible in some way for the differential misexpression of certain miRNAs. A genomewide miRNA expression profiling study provided evidence that the miRNA signatures of three human heart pathologies—aortic stenosis, dilated cardiomyopathy and ischemic cardiomyopathy—are distinct, with 43 out of 87 tested miRNAs differentially expressed in at least one of these disease groups. A subsequent study also demonstrated differences and commonalities in the expression of 33 miRNAs in ischemic cardiomyopathy and idiopathic dilated cardiomyopathy. Apart from etiology and pathogenesis, disease progression might affect miRome.

REFERENCES

Carè A, Catalucci D, Felicetti F, Bonci D, Addario A, Gallo P, Bang ML, Segnalini P, Gu Y, Dalton ND, Elia L, Latronico MV, Høydal M, Autore C, Russo MA, Dorn GW, Ellingsen O, Ruiz-Lozano P, Peterson KL, Croce CM, Peschle C, Condorelli G. (2007) MicroRNA-133 controls cardiac hypertrophy. Nat Med 13:613–618.

Chen JF, Mandel EM, Thomson JM, Wu Q, Callis TE, Hammond SM, Conlon FL, Wang DZ. (2006) The role of microRNA-1 and microRNA-133 in skeletal muscle proliferation and differentiation. Nat Genet 38:228–233.

Cheng Y, Ji R, Yue J, Yang J, Liu X, Chen H, Dean DB, Zhang C. (2007) MicroRNAs are aberrantly expressed in hypertrophic heart. Do they play a role in cardiac hypertrophy? Am J Pathol 170:1831–1840.

Divakaran V, Mann DL. (2008) The emerging role of microRNAs in cardiac remodeling and heart failure. Circ Res 103:1072–1083.

Ikeda S, Kong SW, Lu J, Bisping E, Zhang H, Allen PD, Golub RD, Pieske B, Pu WT. (2007) Altered microRNA expression in human heart disease. Physiol Genomics 31:367–373.

Ji R, Cheng Y, Yue J, Yang J, Liu X, Chen H, Dean DB, Zhang C. (2007) MicroRNA expression signature and antisense-mediated depletion reveal an essential role of MicroRNA in vascular neointimal lesion formation. Circ Res 100: 1579–1588.

Lagos-Quintana M, Rauhut R, Yalcin A, Meyer J, Lendeckel W, Tuschl T. (2002) Identification of tissue-specific miRNAs from mouse. Curr Biol 12: 735–739.

Latronico MVG, Catalucci D, Condorelli G. (2007) Emerging Role of MicroRNAs in Cardiovascular Biology. Circ Res 101:1225–1236

Liang Y, Ridzon D, Wong L, Chen C. (2007) Characterization of microRNA expression profiles in normal human tissues. BMC Genomics 8:166.

Luo X, Zhang H, Xiao J, Wang Z. (2010) Regulation of human cardiac ion channel genes by microRNAs: Theoretical perspective and pathophysiological implications. *Cell Physiol Biochem* (in press).

Miranda KC, Huynh T, Tay Y, Ang YS, Tam WL, Thomson AM, Lim B, Rigoutsos I. (2006) A pattern-based method for the identification of MicroRNA binding sites and their corresponding heteroduplexes. Cell 126:1203–1217.

Morkin E. (2000) Control of cardiac myosin heavy chain gene expression. Microsc Res Tech 50:522–531.

Rao PK, Kumar RM, Farkhondeh M, Baskerville S, Lodish HF. (2006) Myogenic factors that regulate expression of muscle-specific microRNAs. Proc Natl Acad Sci USA 103:8721–8726.

Sucharov C, Bristow MR, Port JD. (2008) miRNA expression in the failing human heart: functional correlates. J Mol Cell Cardiol 45:185–192.

Thum T, Galuppo P, Wolf C, Fiedler J, Kneitz S, van Laake LW, Doevendans PA, Mummery CL, Borlak J, Haverich A, Gross C, Engelhardt S, Ertl G, Bauersachs J. (2007) MicroRNAs in the human heart: a clue to fetal gene reprogramming in heart failure. Circulation 116:258–267.

Rosenkranz-Weiss P, Tomek RJ, Mathew J, Eghbali M. (1994) Gender-specific differences in expression of mRNAs for functional and structural proteins in rat ventricular myocardium. J Mol Cell Cardiol 26:261–270.

van Rooij E, Quiat D, Johnson BA, Sutherland LB, Qi X, Richardson JA, Kelm RJ Jr, Olson EN. (2009) A family of microRNAs encoded by myosin genes governs myosin expression and muscle performance. Dev Cell 17:662–673.

van Rooij E, Sutherland LB, Qi X, Richardson JA, Hill J, Olson EN. (2007) Control of stress-dependent cardiac growth and gene expression by a microRNA. Science 316:575–579.

van Rooij E, Sutherland LB, Liu N, Williams AH, McAnally J, Gerard RD, Richardson JA, Olson EN. (2006) A signature pattern of stress-responsive microRNAs that can evoke cardiac hypertrophy and heart failure. Proc Natl Acad Sci USA 103:18255–18260.

Wang Z, Yang B. (2010) microRNA expression detection methods. Springer-Verlag, New York, USA.

Weiss A, Leinwand LA. (1996) The mammalian myosin heavy chain gene family. Annu Rev Cell Dev Biol 12:417–439.

Zhao Y, Samal E, Srivastava D. (2005) Serum response factor regulates a muscle-specific microRNA that targets Hand2 during cardiogenesis. Nature 436:214–220.

miRNAs in Cardiac Development

Abstract: This chapter aims to summarize the available data on regulation of cardiac development and stem cell differentiation by miRNAs. Heart malformations occur in as high as 1% of newborns, presenting a significant clinical problem in our modern world. The first functional organ in the embryo is the heart and cardiovascular system and the heart is susceptible to congenital defects more than any other organ. Both intrinsic and extrinsic factors determine the development of the cardiovascular system. miRNA was initially described as being fundamental for developmental biology first in nematode worms and then in phylogenically more advanced organisms. Many defects of the miRNA machinery are incompatible with correct and/or continued development. On the other hand, pluripotency and cellular differentiation are intricate biological processes that are coordinately regulated by a complex set of factors and epigenetic regulators. As in other tissues, a distinct set of miRNAs is specifically expressed in pluripotent embryonic stem cells. This chapter describes the involvement of miRNAs in normal cardiac development, in congenital heart disease and Down syndrome, and in determining stem cell fate. In particular, the roles of miR-1, miR-133, miR-130a and miR-138 in cardiac development are described as these miRNAs have been experimentally studied in detail.

INTRODUCTION

The first functional organ in the embryo is the heart and cardiovascular system. Both intrinsic and extrinsic factors determine the development of the cardiovascular system. Many genes and cell populations are involved in cardiac-specific mechanisms, such as cardiogenesis, looping, septation, valve formation, pump function, and hemodynamics. More generic processes, such as epithelium-to-mesenchyme transformation or EMT (in the endocardium to form the endocardial cushions, and in the epicardium to form epicardium-derived cells), cell migration (neural crest cells towards both the arterial and the venous pole), ion channel function (central and peripheral conduction system), and apoptosis (neural crest cells, myocardium) play important roles as well. As the cells' transduction machinery is composed of many interlocking pathways, the above-mentioned mechanisms usually interlock and cannot be separated easily.

Cardiogenesis takes place via a complex series of steps: (1) Determination of mesoderm- and neural crest cells for heart formation; (2) Growth and differentiation processes to become cardiomyocytes; and (3) Migration and transformation processes in order to form the heart. The cardiogenic plate is formed by a collection of mesoderm cells in the most anterior part of the embryo.

Heart malformations occur in as high as 1% of newborns, presenting a significant clinical problem in our modern world. In nearly 50% of prenatally lethal mouse knockouts, the ardiovascular/hemopoietic system is affected, underscoring the importance for embryo survival. It is small wonder that the complexity of heart development often results in malformations, the largest group of congenital anomalies. Many nonlethal cardiovascular anomalies are encountered that nevertheless need clinical intervention to improve quality of life, providing a huge cost factor for society.

Recent research has uncovered a new layer of regulation of cardiac development: miRNAs are essential for the correct development of animals, regulating the formation of tissues and organs. Indeed, miRNA was initially described as being fundamental for developmental biology first in nematode worms and then in phylogenically more advanced organisms. Not surprisingly, many defects of the miRNA machinery are incompatible with correct and/or continued development. The heart is susceptible to congenital defects more than any other organ, and evidence has been accumulating indicating that miRNAs contribute actively to these is still unknown. A number of reports describing the importance of miRNAs for cardiogenesis in mice [Callis *et al.*, 2007] have shown that misexpression can produce defects.

miRNAs AND CARDIAC DEVELOPMENTAL DISORDERS

Zhao *et al* [2007] engineered mice that lacked the miRNA processing enzyme Dicer specifically in heart tissue using the Cre-Lox technology. In this model, the Cre recombinase is under control of the endogenous Nkx2.5 regulatory region driving expression in cardiac progenitors by embryonic day 8.5. Embryos died early (by embryonic day 12.5) from cardiac failure because of a variety of developmental defects, including pericardial edema and underdevelopment of the ventricular myocardium. To further analyze the role of miRNAs in cardiac development, Zhao *et al* [2005] focused on miR-1, a highly abundant class of miRNAs in mammalian heart. The miR-1 subfamily consists of two closely related miRNAs encoded by distinct genes that share near complete identity and are designed miR-1-1 and miR-1-2. Targeted deletion of miR-1-2 using homologous recombination in mouse embryonic stem cells, induced death during the embryonic development in 50% of homozygous mice. Remarkably, a redundant miR-1-1 locus did not compensate for loss of miR-1-2, suggesting that miR-1 dosage is important for heart development and function because major ventricular septum defects were observed. In a previous work Zhao *et al* [2005] reported that miR-1-1 and miR-1-2 are expressed in a chamber-specific manner during cardiogenesis (ubiquitous in the heart and specific for the ventricle, respectively) and are activated during the period of cardiomyocyte differentiation. Both genes are direct targets of serum response factor (SRF) and its potent co-activator myocardin. They negatively regulate ventricular cardiomyocyte proliferation through targeting of the mRNA of the Hand2 gene, a cardiac transcription factor crucial for cardiomyocyte development and morphogenesis. Mi-R1 titrates the effects of critical cardiac regulatory proteins (Hrt2/Hey2, Hand1, Hand2, Gata6, etc.) to control the balance between differentiation and proliferation of progenitor cells during cardiogenesis. These findings may have implications for the development of congenital heart disease, because malfunctioning miRNAs could adversely affect cardiac development. They could also give some insight into the molecular mechanisms controlling specific cell-fate decisions, possibly clarifying the processes by which a progenitor or a stem cell evolves into a cardiomyocyte. This might lead to new therapeutic approaches for the regeneration of damaged heart muscle [Srivastava, 2006+].

Mutations in Dicer or Argonaute cause severe developmental phenotypes in several organisms [Grishok *et al.*, 2001; Knight & Bass, 2001; Wienholds *et al.*, 2003; Bernstein *et al.*, 2003; Giraldez *et al.*, 2005; Hatfield *et al.*, 2005; Deshpande *et al.*, 2005; Cox *et al.*, 1998]. In mice, a Dicer knockout is lethal at day E7.5, pointing to early and essential roles for miRNAs in development, and the inability of Dicer null embryonic stem cells to proliferate suggests a role in stem cell maintenance [Bernstein *et al.*, 2003]. Block of miRNA biogenesis by tissue-specific deletion of Dicer causes lethality of embryos due to defects in cardiogenesis. When Dicer was efficiently deleted in the heart, death of the embryos by E12.5 due to cardiac failure was consistently observed [Zhao *et al.*, 2007]. Dicer-deficient embryos developed cardiac failure due to pericardial edema and underdevelopment of the ventricular myocardium. These phenotypes are in agreement with the defects of heart development in zebrafish embryos devoid of Dicer function [Giraldez *et al.*, 2005].

miRNAs in Normal Cardiac Development

Tissue development depends on the right spatiotemporal expression of relevant protein-coding genes which are critically regulated in their expression by miRNAs. Indeed, certain spatiotemporal patterns of expression of particular miRNAs have been implicated in the development of tissues/organs. Cell- and/or tissue-specific modulation of miRNA expression levels has been demonstrated to determine the fine-tuning of specific targets that need to be activated or silenced at a certain stage of development [Farh *et al.*, 2005]. Role of miRNAs in cardiac development and the related disorders have been intensively studied. It is now commonly recognized that miRNAs are a critical player in orchestrating morphogenesis as well as in early embryonic patterning processes. However, relatively few miRNAs have been identified to regulate cardiac development.

miR-1 and miR-133

Studies using both "gain-of-function" and "loss-of-function" approaches point to the roles of miR-1 and miR-133 in modulating cardiogenesis and in maintenance of muscle-gene expression. These muscle-specific miRNAs are expressed in the early stages during cardiogenesis. Microarray analysis revealed an increased expression of miR-1 and miR-133 in developing mouse hearts from embryonic day E12.5 through to at least E18.5, indicating a requirement of these miRNAs for cardiac development [Chen *et al.*, 2006]. Several lines of evidence have been obtained in support of this notion [Zhao *et al.*, 2005; Kwon *et al.*, 2005; Zhao *et al.*, 2007; Niu *et al.*, 2007]. Loss-of function of miR-1 in *Drosophila* resulted in embryonic/larval lethality with most of the mutant flies displaying altered sarcomeric gene expression and, in a subset of embryos, an increased number of undifferentiated muscle progenitors [Kwon *et al.*, 2005]. In mouse, miR-1 is responsible for the inhibition of cardiomyocyte progenitor proliferation via inhibition of translation of Hand2 [Zhao *et al.*, 2005], a transcription factor known to regulate ventricular cardiomyocyte expansion [Srivastava *et al.*, 1997]. Many of the embryos from miR-1-2 knockout mice demonstrated ventricular septal defects in a subset that suffer early lethality. The adult miR-1-2-deficient mice had thickened chamber walls attributable to hyperplasia of the heart [Zhao *et al.*, 2007]. In contrast, miR-1 gain-of-function led to 100% embryonic fly lethality because of disrupted patterning of cardiac and skeletal muscle with insufficient numbers of cardioblasts. miR-1 controls heart development in mice by regulation of the cardiac transcription factor Hand2 [Zhao *et al.*, 2005]. Overexpression of miR-1 in a transgenic mouse model resulted in a phenotype characterized by thin-walled ventricles, attributable to premature differentiation and early withdrawal of cardiomyocytes from the cell cycle [Zhao *et al.*, 2007]. Altogether, these results reveal a tight control of spatiotemporal miR-1 expression for proper cardiac and/or skeletal muscle development.

Genetic engineering in *Drosophila* demonstrated that *miR-1* may "fine-tune" Notch ligand Delta that is critically involved in differentiation of cardiac and somatic muscle progenitors and targets a pathway required for progenitor cell specification and asymmetric cell division [Kwon *et al.*, 2005]. In *Xenopus laevis*, injection of either miR-1 at one-cell stage results in complete absence of cardiac tissue [Chen *et al.*, 2006]. On the other hand, introduction of miR-133 allows cardiac tissue to form, but the tissue is disorganized and does not lead to chamber formation. The authors of this study proposed that both miR-1 and miR-133 are required for proper heart development and they may have distinct roles with miR-1 promoting myogenesis while miR-133 enhancing myoblast proliferation.

In a separate study to dissect the role of the members of the miR-133 subfamily [Liu *et al.*, 2008], it was found that mice lacking either miR-133a-1 or miR-133a-2 are normal, whereas deletion of both miRNAs causes lethal ventricular-septal defects in approximately half of double-mutant embryos or neonates; miR-133a double-mutant mice that survive to adulthood succumb to dilated cardiomyopathy and heart failure. The absence of miR-133a expression results in ectopic expression of smooth muscle genes in the heart and aberrant cardiomyocyte proliferation. These abnormalities can be attributed, at least in part, to elevated expression of SRF and cyclin D2, which are targets for repression by miR-133a. These findings reveal essential and redundant roles for miR-133a-1 and miR-133a-2 in orchestrating cardiac development, gene expression, and function and point to these miRNAs as critical components of an SRF-dependent myogenic transcriptional circuit.

miR-130a

FOG-2 (also known as zfpm2) is a transcriptional co-factor that is critical for cardiac development. In one report, it was demonstrated that FOG-2 expression is controlled at the translational level by miR-130a [Kim *et al.*, 2009]. There is a conserved region in the FOG-2 3'UTR predicted to be a target for miR-130a. The authors established FOG-2 as a cognate target for miR-130a in NIH 3T3 fibroblasts, and observed a 3.3-fold increase in translational efficiency when the miR-130a target site in FOG-2 was disrupted. Knockdown of miR-130a in fibroblasts resulted in a 3.6-fold increase in translational efficiency. They further demonstrated that cardiomyocytes express miR-130a which can attenuate translation of the FOG-2 mRNA. They generated transgenic mice with cardiomyocyte over-expression of miR-130a. In the hearts of these mice, FOG-2 protein levels were reduced by ~80%. Histological analysis of transgenic embryos revealed ventricular wall hypoplasia and ventricular septal defects, similar to that seen in FOG-2 deficient hearts [Svensson *et al.*, 2000]. These results demonstrate the importance of miR-130a for the regulation of FOG-2 protein expression and suggest that miR-130a may also play a role in the regulation of cardiac development, in addition to the muscle-specific miRNAs miR-1 and miR-133 as demonstrated by other groups.

miR-138

Organ patterning during embryonic development requires precise temporal and spatial regulation of protein activity; miR-138 was found to play a role in organ patterning [Morton *et al.*, 2008]. miR-138 is expressed in specific domains in the zebrafish heart and is required to establish appropriate chamber-specific gene expression patterns [Morton *et al.*, 2008]. Disruption of miR-138 function led to ventricular expansion of gene expression normally restricted to the atrio-ventricular valve region and, ultimately, to disrupted ventricular cardiomyocyte morphology and cardiac function. Temporal-specific knockdown of miR-138 by antagomiRs showed miR-138 function was required during a discrete developmental window, 24-34 h post-fertilization (hpf). miR-138 functioned partially by repressing the retinoic acid synthesis enzyme, aldchyde dehydrogenase-1a2, in the ventricle. This activity was complemented by miR-138-mediated ventricular repression of the gene encoding versican (cspg2), which was positively regulated by retinoic-acid signaling. These findings indicate that miR-138 helps establish discrete domains of gene expression during cardiac morphogenesis by targeting multiple members of a common pathway.

miRNAs in Congenital Heart Disease

Congenital heart disease (CHD) is the leading cause of infant morbidity in industrialized countries, with an incidence of ~2% to 8% of live births [Hoffman & Kaplan, 2002]. A genetic component for CHD is implicated by their familial aggregation and by studies showing an association of CHD with inherited microdeletion and mutation syndromes [Petrij *et al.*, 1995; Merscher *et al.*, 2001]. Hox gene clusters play an important role during cardiac septation to valve formation in different species. HOX genes encode a highly-conserved family of transcription factors and have fundamental roles in morphogenesis [Krumlauf, 1994]. miR-196a was identified as an upstream regulator of Hoxb8 and Sonic hedgehog (Shh) *in vivo* in the context of limb development [Hornstein *et al.*, 2005]; this finding suggests a possible role in heart development and CHD, because Shh signaling is required throughout cardiac septation to valve formation [Goddeeris *et al.*, 2007, 2008]. Yekta *et al.* [2004] first reported that miR-196a is expressed from HOX gene clusters loci in mammals and that HOX genes in turn are targets of miR-196a. miR-196a directed Hoxb8 RNA cleavage products have been detected in total RNA extracted from E11.5 (ventricular septation) mouse embryos using a sensor transgene for the Hox complex-embedded miR-196a [Mansfield *et al.*, 2004]. Furthermore, in fetal human heart samples (gestational age of 12–14 weeks), miR-196a has been found to be 1.88-fold (P50.034) upregulated compared to healthy adult heart samples [Thum *et al.*, 2007].

A genetic variant of rs11614913 in the miR-196a2 sequence could alter mature miR-196a expression and target mRNA binding [Hu *et al.*, 2008]. The same group further conducted a three-stage case–control study of CHD in Chinese population to test their hypothesis by genotyping miR-196a2 rs11614913 and three other pre-mina SNPs (miR-146a rs2910164, miR-149 rs2292832, and miR-499 rs3746444) in 1,324 CHD cases and 1,783 non-CHD controls [Xu *et al.*, 2009]. They found that rs11614913 CC was associated with a significantly increased risk of CHD in all three stages combined. In a genotype–phenotype correlation analysis using 29 cardiac tissue samples of CHD, rs11614913 CC was associated with significantly increased mature miR-196a expression. *In vitro* binding assays further revealed that the rs11614913 variant affects HOXB8 binding to mature miR-196a.

miRNAs in Down Syndrome

Down syndrome (DS) or Trisomy 21 is a result of the presence of an extra copy of all or part of human chromosome 21 (Hsa21) [LeJeune *et al.*, 1959]. DS is the most frequent survivable congenital chromosomal abnormality occurring in approximately 750–1000 live births [Hook *et al.*, 1983]. Overexpression of the genes on Hsa21 by 50% in many tissues is thought to initiate the DS phenotypes, however, there is currently no explanation for how this relatively small increase in transcript levels results in any specific feature of DS [Mao *et al.*, 2003; Gardiner & Costa, 2006]. Bioinformatic analyses demonstrate that human chromosome 21 (Hsa21) harbors five miRNA genes; miR-99a, let-7c, miR-125b-2, miR-155, and miR-802. MiRNA expression profiling, miRNA RT-PCR, and miRNA in situ hybridization experiments demonstrate that these miRNAs are overexpressed in fetal brain and heart specimens from individuals with DS when compared with age- and sex-matched controls [Kuhn *et al.*, 2008]. This finding raised a possibility that trisomic 21 gene dosage overexpression of Hsa21-derived miRNAs results in the decreased expression of specific target proteins and contribute, in part, to features of the neuronal and cardiac DS phenotype. Further studies are needed to elucidate the relevant target genes and acquire evidence for cause-effect relationship between the miRNAs and DS.

miRNAs AND STEM CELL FATE

Pluripotency and cellular differentiation are intricate biological processes that are coordinately regulated by a complex set of factors and epigenetic regulators. Human pluripotent stem cell lines can be generated from surplus fertilized eggs or, as demonstrated more recently, from the reprogramming of somatic cells. Cell fate decisions of pluripotent stem cells are dictated by activation and repression of lineage-specific genes. Numerous signaling and transcriptional networks progressively narrow and specify the potential of pluripotent stem cells. Standardized culture conditions for the long-term maintenance and propagation of undifferentiated human pluripotent stem cells have been developed for research. The differentiation of human pluripotent stem cells may enable the generation of large quantities of specialized cells that can be used as *in vitro* tools for drug development, as well as for future applications in regenerative medicine. However, most of the currently used differentiation protocols yield inefficient stem cell quantities and low purity of the final cell preparations. The discovery of miRNAs and their role as important transcriptional regulators may provide a new means of manipulating stem cell fate.

As in other tissues, a distinct set of miRNAs is specifically expressed in pluripotent embryonic stem cells (ESC) but not in differentiated embryoid bodies, suggesting a role of miRNAs in stem cell self-renewal [Houbaviy *et al.*, 2003]. Dicer-null ESCs lack mature miRNAs and thus fail to differentiate into the three germlayers [Kanellopoulou *et al.*, 2005; Murchison *et al.*, 2005]. In Dicer-null mouse embryos, the pool of pluripotent stem cells in the inner-cell mass of the blastocyst is diminished [Bernstein *et al.*, 2003].

The miRNAs that are significantly upregulated during the differentiation of embryonic stem cells to cardiomyocytes are miR-1, miR-18, miR-20, miR-23b, miR-24, miR-26a, miR-30c, miR-133, miR-143, miR-182, miR-183, miR-200a/b, miR-292-3p, miR-293, miR-295 and miR-335 in mice [Srivastava *et al.*, 1997], and miR-1, miR-20, miR-21, miR-26a, miR-92, miR-127, miR-129, miR-130a, miR-199b, miR-200a, miR-335 and miR-424 in humans [Ivey *et al.*, 2008]. A considerable number of miRNAs upregulated in differentiating cardiomyocytes are also enriched in human fetal heart tissue [Thum *et al.*, 2007]. The number of detectable miRNAs increases rapidly in tissues derived from all three germ layers (endoderm, ectoderm, and mesoderm). An analysis of the miRNAs expressed in undifferentiated mouse embryonic stem cells and differentiating cardiomyocytes was recently published [Lakshmipathy *et al.*, 2007].

Mouse Pluripotent Embryonic Stem Cell

Pluripotent embryonic stem cells (ESC) are capable of differentiating to all possible cell types. This unique property promises future medical applications, but to fulfill this potential it will be necessary to control ESC differentiation. Mouse ESCs lacking miRNAs are unable to downregulate pluripotency markers after induction of differentiation and retain the ability to produce ESC colonies [Kloosterman & Plasterk, 2006]. Ivey and colleagues demonstrate a new role for miRNAs in regulating ESC differentiation [Ivey *et al.*, 2008]. They found that miR-1 and miR-133 direct mesoderm formation and regulate differentiation to cardiac muscle by suppressing gene expression of alternative lineages. They first identified 9 miRNAs that are enriched in cardiac progenitors derived from *in vitro*-differentiated mouse ESC cells and undetectable in undifferentiated ESC, among which miR-1 and miR-133 have been known to be of paramount importance for heart development and physiology [Chen *et al.*, 2006; Zhao *et al.*, 2007]. Forced expression of either of these miRNAs by lentiviral vectors was insufficient to drive ES cell differentiation. However, during embryoid body formation, expression of either miR-1 or miR-133 led to dramatically increased levels of early mesoderm markers, due to an increased number of expressing cells as well as increased levels per cell, suggesting that both miRNAs promote mesoderm formation. However, whereas miR-1 also promoted further differentiation to cardiac and skeletal muscle, expression of miR-133 had an inhibitory effect on myogenesis. Therefore, these *in vitro* ESC differentiation assays largely recapitulate several aspects of miR-1 and miR-133 function *in vivo* [Chen *et al.*, 2006; Zhao *et al.*, 2007]. The authors further compared gene expression profiles of wild-type and miR-1 and miR-133-expressing cells by microarray analyses. Expression of markers associated with endoderm specification and differentiation was significantly lower in the miRNA-expressing cells. Compared to control teratomas, those derived from miR-1- and miR-133-expressing ES cells had more neuronal progenitor cells, but were dramatically depleted of differentiated neuronal cells. The physiological relevance of promoting neuronal, but not endoderm, progenitor cells by these muscle-specific miRNAs is not clear, but it certainly warrants further investigation. Nevertheless, it is clear that both miRNAs suppress cell differentiation into nonmuscle lineages. And both miRNAs seemed to have comparable functions in human ES cells.

In a separate study, Takaya *et al* [2009] obtained different observations from Ivey *et al* [2008] on the role of miR-1 and miR-133 in spontaneous myocardial differentiation. They found that the levels of miR-1 and miR-133 in mouse ESCs were increased during spontaneous differentiation by 2-dimensional culture, but reduced during forced myocardial differentiation by a histone deacetylase inhibitor, trichostatin A. Forced expression of miR-1 or miR-133, but not that of miR-143, by lentiviral infection reduced the expression of a cardiac-specific gene, Nkx2.5, during differentiation of ESCs. In addition, miR-1 also inhibited α-myosin heavy chain expression. The results of luciferase assays revealed that miR-1 recognizes and targets the 3' untranslated region of cyclin-dependent kinase-9 (Cdk9) in ESCs. Overexpression of miR-1 decreased the protein amounts of Cdk9 without affecting the mRNA levels, indicating that miR-1 post-transcriptionally inhibits Cdk9 translation. Clearly, miR-1 and miR-133 may play significant roles in the myocardial differentiation of mouse ES cells, and Cdk9 may be involved in this process as a target of miR-1.

Human Fetal Cardiomyocyte Progenitor Cells

Cardiomyocyte progenitor cells (CMPCs) are a promising cell population to improve regeneration of the injured myocardium, through enhancing the intrinsic capacity of the heart to regenerate itself and/or replace the damaged tissue by cell transplantation. They are easily expanded and efficiently differentiated into beating cardiomyocytes. Sluijter *et al* [2010] investigated miRNAs involved in proliferation/differentiation of the human CMPCs *in vitro*. Human fetal CMPCs were isolated, cultured, and efficiently differentiated into beating cardiomyocytes. miRNA expression profiling demonstrated that cardiac-specific miR-1 and miR-499 were highly upregulated in differentiated cells. Transient transfection of miR-1 and -499 in CMPC reduced proliferation rate by 25% and 15%, respectively, and enhanced differentiation into cardiomyocytes in human CMPCs and embryonic stem cells, likely via the repression of histone deacetylase 4 or Sox6. Histone deacetylase 4 and Sox6 protein levels were reduced, and siRNA-mediated knockdown of Sox6 strongly induced myogenic differentiation. Evidently, by modulating miR-1 and miR-499 expression levels, human CMPC function can be altered and differentiation directed, thereby enhancing cardiomyogenic differentiation.

Bone Marrow-derived Mesenchymal Stem Cells

Won Kim *et al* [2009] investigated the role of miRNAs in ischemic preconditioning (IP) of bone marrow-derived mesenchymal stem cells (MSCs). IP of MSCs with two cycles of 30-min ischemia/reoxygenation (I/R) supported their survival under subsequent longer exposure to anoxia and following engraftment in the infarcted heart. IP significantly reduced apoptosis in MSCs through activation of Akt and ERK1/2 and nuclear translocation of hypoxia-inducible factor-1alpha (HIF-1α). They then observed concomitant induction of miR-210 in the preconditioned MSCs. Inhibition of HIF-1α or downregulation of miR-210 abrogated the cytoprotective effects of IP. Further, engraftment with miR-210 improved stem cell survival in a rat model of acute myocardial infarction. Notably, multiple I/R cycles more effectively regulated the miR-210 and hence promoted MSC survival compared with single-cycle hypoxia of an equal duration. They then identified FLICE-associated huge protein (FLASH)/caspase-8-associated protein-2 in MSCs as the target gene of miR-210. Upregulation of these genes through miR-210 knockdown in MSCs resulted in increased cell apoptosis. These data demonstrated that cytoprotection afforded by IP was regulated by miR-210 induction via FLASH/ caspase-8-associated protein-2 repression. These results highlighted that IP by multiple short episodes of I/R is a novel strategy to promote stem cell survival.

Multipotent Murine Cardiac Progenitor Cells

Evidence indicates that a multipotent cardiac progenitor pool exists that can differentiate into cardiac myocytes, vascular smooth muscle cells (VSMCs) or endothelial cells [Kattman *et al.*, 2006]. Among these cell types, VSMCs are uniquely plastic, as they can oscillate between a proliferative or a quiescent, more differentiated state [Ross, 1993]. This plasticity contributes to many human vascular diseases, including atherosclerosis [Owens *et al.*, 2004; Yoshida & Owens, 2004]. Srivastava's group has reported that miR-143 is the most enriched miRNA during differentiation of mouse embryonic stem cells (ESCs) into multipotent cardiac progenitors [Ivey *et al.*, 2008]. The same group in a more recent study [Cordes *et al.*, 2009] showed that miR-145 and miR-143 were co-transcribed in multipotent murine cardiac progenitors before becoming localized to smooth muscle cells, including neural crest stem-cell-derived vascular smooth muscle cells. miR-145 and miR-143 were direct transcriptional targets of serum response factor (SRF), myocardin and NK2 transcription factor related, locus 5 (Nkx2-5) and were found downregulated in injured or atherosclerotic vessels containing proliferating, less differentiated smooth muscle cells. miR-145 was necessary for myocardin-induced reprogramming of adult fibroblasts into smooth muscle cells and sufficient to induce differentiation of multipotent neural crest stem cells into vascular smooth muscle. Furthermore, miR-145 and miR-143 cooperatively targeted a network of transcription factors, including Klf4 (Kruppel-like factor 4), myocardin and Elk-1 (ELK1, member of ETS oncogene family), to promote differentiation and repress proliferation of smooth muscle cells. These findings demonstrate that miR-145 can direct the smooth muscle fate and that miR-145 and miR-143 function to regulate the quiescent versus proliferative phenotype of smooth muscle cells and provide evidence for a single miRNA that can efficiently differentiate multipotent stem cells into a specific lineage or regulate direct reprogramming of cells into an alternative cell fate.

REFERENCES

Allerson CR, Sioufi N, Jarres R, Prakash TP, Naik N, Berdeja A, Wanders L, Griffey RH, Swayze EE, Bhat B. (2005) Fully 2'-modified oligonucleotide duplexes with improved *in vitro* potency and stability compared to unmodified small interfering RNA. J Med Chem 48:901–904.

Altmann KH, Dean NM, Fabbro D, Freier SM, Geiger T, Haener R, Huesken D, Martin P, Monia BP, Muller M, Natt F, Nicklin P, Phillips J, Pieles U, Sasmor H, Moser H. (1996) Second generation of antisense oligonucleotides. From nuclease resistance to biological efficacy in animals. Chimia 50:168–176.

Amarzguioui M, Holen T, Babaie E, Prydz H. (2003) Tolerance for mutations and chemical modifications in a siRNA. Nucleic Acids Res 31:589–595.

Bernstein E, Kim SY, Carmell MA, Murchison EP, Alcorn H, Li MZ, Mills AA, Elledge SJ, Anderson KV, Hannon GJ. (2003) Dicer is essential for mouse development. Nat Genet 35:215–217.

Boden D, Pusch O, Silbermann R, Lee F, Tucker L, Ramratnam B. (2004) Enhanced gene silencing of HIV-1 specific siRNA using microRNA designed hairpins. Nucleic Acids Res 32:1154–1158.

Braasch DA, Jensen S, Liu Y, Kaur K, Arar K, White MA, Corey DR. (2003) RNA interference in mammalian cells by chemically-modified RNA. Biochemistry 42:7967–7975.

Braasch DA, Paroo Z, Constantinescu A, Ren G, Oz OK, Mason RP, Corey DR. (2004) Biodistribution of phosphodiester and phosphorothioate siRNA. Bioorg Med Chem Lett 14:1139–1143.

Brummelkamp TR, Bernards R, Agami R. (2002a). Stable suppression of tumorigenicity by virus-mediated RNA interference. Cancer Cell 2:243–247.

Brummelkamp TR, Bernards R, Agami R. (2002b) A system for stable expression of short interfering RNAs in mammalian cells. Science 296:550–553.

Buchschacher Jr G.L, Wong-Staal F. (2000) Development of lentiviral vectors for gene therapy for human diseases. Blood 95:2499–2504.

Callis TE, Chen JF, Wang DZ. (2007) MicroRNAs in skeletal and cardiac muscle development. DNA Cell Biol 26:219–225.

Chen JF, Mandel EM, Thomson JM, Wu Q, Callis TE, Hammond SM, Conlon FL, Wang DZ. (2006) The role of microRNA-1 and microRNA-133 in skeletal muscle proliferation and differentiation. Nat Genet 38:228–233.

Chen C, Ridzon DA, Broomer AJ, Zhou Z, Lee DH, Nguyen JT, Barbisin M, Xu NL, Mahuvakar VR, Andersen MR, Lao KQ, Livak KJ, Guegler KJ. (2005) Real-time quantification of microRNAs by stem-loop RT-PCR. Nucleic Acids Res 33:e179.

Chiu YL, Rana TM. (2003) siRNA function in RNAi: a chemical modification analysis. RNA 9:1034–1048.

Chang TC, Wentzel EA, Kent OA, Ramachandran K, Mullendore M, Lee KH, Feldmann G, Yamakuchi M, Ferlito M, Lowenstein CJ, Arking DE, Beer MA, Maitra A, Mendell JT. (2007) Transactivation of miR-34a by p53 broadly influences gene expression and promotes apoptosis. Mol Cell 26:745–752.

Chung KH, Hart CC, Al-Bassam S, Avery A, Taylor J, Patel PD, Vojtek AB, Turner DL. (2006) Polycistronic RNA polymerase II expression vectors for RNA interference based on BIC/miR-155. Nucleic Acids Res 34:e53.

Clop A, Marcq F, Takeda H, Pirottin D, Tordoir X, Bibe B, Bouix J, Caiment F, Elsen JM, Eychenne F, Larzul C, Laville E, Meish F, Milenkovic D, Tobin J, Charlier J, Georges M. (2006) A mutation creating a potential illegitimate microRNA target site in the myostatin gene affects muscularity in sheep. Nat Genet 38:813–818.

Cordes KR, Sheehy NT, White MP, Berry EC, Morton SU, Muth AN, Lee TH, Miano JM, Ivey KN, Srivastava D. (2009) miR-145 and miR-143 regulate smooth muscle cell fate and plasticity. Nature 460:705–710.

Corney DC, Flesken-Nikitin A, Godwin AK, Wang W, Nikitin AY. (2007) MicroRNA-34b and MicroRNA-34c are targets of p53 and cooperate in control of cell proliferation and adhesion-independent growth. Cancer Res 67:8433–8438.

Czauderna F, Fechtner M, Dames S, Aygun H, Klippel A, Pronk GJ, Giese K, Kaufmann J. (2003) Structural variations and stabilising modifications of synthetic siRNAs in mammalian cells. Nucleic Acids Res 31:2705–2716.

Deo M, Yu JY, Chung KH, Tippens M, Turner DL. (2006) Detection of mammalian microRNA expression by in situ hybridization with RNA oligonucleotides. Dev Dyn 235:2538–2548.

Eisenberg I, Eran A, Nishino I, Moggio M, Lamperti C, Amato AA, Lidov HG, Kang PB, North KN, Mitrani-Rosenbaum S, Flanigan KM, Neely LA, Whitney D, Beggs AH, Kohane IS, Kunkel LM. (2007) Distinctive patterns of microRNA in primary muscular disorders. Proc Natl Acad Sci USA 104:17016–17021.

Gardiner K, Costa AC. (2006) The proteins of human chromosome 21. Am J Med Genet 142C:196–205.

Ge Q, Filip L, Bai A, Nguyen T, Eisen HN, Chen J. (2004) Inhibition of influenza virus production in virus-infected mice by RNA interference. Proc Natl Acad Sci USA 101:8676–8681.

Goddeeris MM, Schwartz R, Klingensmith J, Meyers EN. (2007) Independent requirements for Hedgehog signaling by both the anterior heart field and neural crest cells for outflow tract development. Development 134:1593–1604.

Goddeeris MM, Rho S, Petiet A, Davenport CL, Johnson GA, Meyers EN, Klingensmith J. (2008) Intracardiac septation requires hedgehog-dependent cellular contributions from outside the heart. Development 135:1887–1895.

Griffiths-Jones S. (2006) miRBase: the microRNA sequence database. Methods Mol Biol342:29–138.

Guglielmelli P, Tozzi L, Pancrazzi A, Bogani C, Antonioli E, Ponziani V, Poli G, Zini R, Ferrari S, Manfredini R, Bosi A, Vannucchi AM; MPD Research Consortium. (2007) MicroRNA expression profile in granulocytes from primary myelofibrosis patients. Exp Hematol 35:1708–1718.

Harborth J, Elbashir SM, Vandenburgh K, Manninga H, Scaringe SA, Weber K, Tuschl T. (2003) Sequence, chemical, and structural variation of small interfering RNAs and short hairpin RNAs and the effect on mammalian gene silencing. Antisense Nucleic Acid Drug Dev 13:83–105.

He L, He X, Lim LP, de Stanchina E, Xuan Z, Liang Y, Xue W, Zender L, Magnus J, Ridzon D, Jackson AL, Linsley PS, Chen C, Lowe SW, Cleary MA, Hannon GJ. (2007) A microRNA component of the p53 tumour suppressor network. Nature 447:1130–1134.

Hoffman JI, Kaplan S. (2002) The incidence of congenital heart disease. J Am Coll Cardiol 39:1890–1900.

Hoke GD, Draper K, Freier SM, Gonzalez C, Driver VB, Zounes MC, Ecker DJ. (1991) Effects of phosphorothioate capping on antisense oligonucleotide stability, hybridization and antiviral efficacy versus herpes simplex virus infection. Nucleic Acids Res 19:5743–5748.

Hook EG, Cross PK, Schreinemachers DM. (1983) Chromosomal abnormality rates at amniocentesis and in live-born infants. J Am Med Assoc 249:2034–2038.

Hornstein E, Mansfield JH, Yekta S, Hu JK, Harfe BD, McManus MT, Baskerville S, Bartel DP, Tabin CJ. (2005) The microRNA miR-196 acts upstream of Hoxb8 and Shh in limb development. Nature 438:671–674.

Houbaviy HB, Murray MF, Sharp PA. (2003) Embryonic stem cell-specific microRNAs. Dev Cell 5:351–358.

Hu Z, Chen J, Tian T, Zhou X, Gu H, Xu L, Zeng Y, Miao R, Jin G, Ma H, Chen Y, Shen H. (2008) Genetic variants of miRNA sequences and non-small cell lung cancer survival. J Clin Invest 118:2600–2608.

Ivey KN, Muth A, Arnold J, King FW, Yeh RF, Fish JE, Hsiao EC, Schwartz RJ, Conklin BR, Bernstein HS, Srivastava D. (2008) MicroRNA regulation of cell lineages in mouse and human embryonic stem cells. Cell Stem Cell 2:219–229.

Izzotti A, Calin GA, Arrigo P, Steele VE, Croce CM, De Flora S. (2009) Downregulation of microRNA expression in the lungs of rats exposed to cigarette smoke. FASEB J 23:806–812.

Kanellopoulou C, Muljo SA, Kung AL, Ganesan S, Drapkin R, Jenuwein T, Livingston DM, Rajewsky K. (2005) Dicer-deficient mouse embryonic stem cells are defective in differentiation and centromeric silencing. Genes Dev 19:489–501.

Kattman SJ, Huber TL, Keller GM. (2006) Multipotent flk-11 cardiovascular progenitor cells give rise to the cardiomyocyte, endothelial, and vascular smooth muscle lineages. Dev Cell 11:723–732.

Kim DH, Behlke MA, Rose SD, Chang MS, Choi S, Rossi JJ. (2005) Synthetic dsRNA Dicer substrates enhance RNAi potency and efficacy. Nat Biotechnol 23:222–226.

Kim GH, Samant SA, Earley JU, Svensson EC. (2009) Translational control of FOG-2 expression in cardiomyocytes by microRNA-130a. PLoS One 4:e6161.

Kloosterman WP, Plasterk RH. (2006) The diverse functions of microRNAs in animal development and disease. Dev Cell 11:441–450.

Kraynack BA, Baker BF. (2006) Small interfering RNAs containing full 2′-Omethylribonucleotide-modified sense strands display Argonaute2/eIF2C2-dependent activity. RNA 12:163–176.

Krichevsky AM, King KS, Donahue CP, Khrapko K, Kosik KS. (2003) A microRNA array reveals extensive regulation of microRNAs during brain development. RNA 9:1274–1281.

Krumlauf R. (1994) Hox genes in vertebrate development. Cell 78:191–201.

Kuhn DE, Nuovo GJ, Martin MM, Malana GE, Pleister AP, Jiang J, Schmittgen TD, Terry AV Jr, Gardiner K, Head E, Feldman DS, Elton TS. (2008) Human chromosome 21-derived miRNAs are overexpressed in Down syndrome brains and hearts. Biochem Biophys Res Commun 370:473–477.

Kwon C, Han Z, Olson EN, Srivastava D. (2005) MicroRNA1 influences cardiac differentiation in Drosophila and regulates Notch signaling. Proc Natl Acad Sci USA 102:18986–18991.

Lakshmipathy U, Love B, Goff LA, Jörnsten R, Graichen R, Hart RP, Chesnut JD. (2007) MicroRNA expression pattern of undifferentiated and differentiated human embryonic stem cells. Stem Cells Dev 16:1003–1016.

Lau N, Lim L, Weinstein E, Bartel DP. An abundant class of tiny RNAs with probable regulatory roles in Caenorhabditis elegans. Science 2001;294:858–862.

Layzer JM, McCaffrey AP, Tanner AK, Huang Z, Kay MA, Sullenger BA. (2004) *In vivo* activity of nuclease-resistant siRNAs. RNA 10:766–771.

LeJeune J, Gautier M, Turpin R. (1959) Study of somatic chromosomes from 9 mongoloid children. Comptes Renus de l' Academic les Sciences 248:1721–1722.

Lodygin D, Tarasov V, Epanchintsev A, Berking C, Knyazeva T, Körner H, Knyazev P, Diebold J, Hermeking H. (2008) Inactivation of miR-34a by aberrant CpG methylation in multiple types of cancer. Cell Cycle 7:2591–2600.

Lee RC, Feinbaum RL, Ambros V. (1993) The *C. elegans* heterochronic gene lin-4 encodes small RNAs with antisense complementarity to lin-14. Cell 75:843–854.

Liu N, Bezprozvannaya S, Williams AH, Qi X, Richardson JA, Bassel-Duby R, Olson EN. (2008) microRNA-133a regulates cardiomyocyte proliferation and suppresses smooth muscle gene expression in the heart. Genes Dev 22:3242–3254.

Luo X, Lin H, Lu Y, Li B, Xiao J, Yang B, Wang Z. (2007) Transcriptional activation by stimulating protein 1 and post-transcriptional repression by muscle-specific microRNAs of I_{Ks}-encoding genes and potential implications in regional heterogeneity of their expressions. J Cell Physiol 212:358–367.

Luo X, Lin H, Pan Z, Xiao J, Zhang Y, Lu Y, Yang B, Wang Z. (2008) Overexpression of Sp1 and downregulation of miR-1/miR-133 activates re-expression of pacemaker channel genes HCN2 and HCN4 in hypertrophic heart. J Biol Chem 283:20045–20052.

Luo X, Xiao J, Lin H, Chen G, Zhang H, Zhang J, Lu Y, Yang B, Wang Z. (2009) Discovery of new functional miRNAs urges miRNA genome studies. Cell (in review).

Mansfield JH, Harfe BD, Nissen R, Obenauer J, Srineel J, Chaudhuri A, Farzan-Kashani R, Zuker M, Pasquinelli AE, Ruvkun G, Sharp PA, Tabin CJ, McManus MT. (2004) MicroRNA-responsive 'sensor' transgenes uncover Hox-like and other developmentally regulated patterns of vertebrate microRNA expression. Nat Genet 36:1079–1083.

Mao R, Zielke CL, Zielke HR, Pevsner J. (2003) Global up-regulation of chromosome 21 gene expression in the developing down syndrome brain. Genomics 81:457–467.

McCaffrey AP, Meuse L, Pham TT, Conklin DS, Hannon GJ, Kay MA. (2002) RNA interference in adult mice. Nature 418:38–39.

Merscher S, Funke B, Epstein JA, Heyer J, Puech A, Lu MM, Xavier RJ, Demay MB, Russell RG, Factor S, Tokooya K, Jore BS, Lopez M, Pandita RK, Lia M, Carrion D, Xu H, Schorle H, Kobler JB, Scambler P, Wynshaw-Boris A, Skoultchi AI, Morrow BE, Kucherlapati R. (2001) TBX1 is responsible for cardiovascular defects in velo-cardio-facial/DiGeorge syndrome. Cell 104:619–629.

Miyagishi M, Taira K. (2002) U6 promoter-driven siRNAs with four uridine 30 overhangs efficiently suppress targeted gene expression in mammalian cells. Nat Biotechnol 20:497–500.

Morrissey DV, Blanchard K, Shaw L, Jensen K, Lockridge JA, Dickinson B, McSwiggen JA, Vargeese C, Bowman K, Shaffer CS, Polisky BA, Zinnen S. (2005a) Activity of stabilized short interfering RNA in a mouse model of hepatitis B virus replication. Hepatology 41:1349–1356.

Morrissey DV, Lockridge J.A, Shaw L, Blanchard K, Jensen K, Breen W, Hartsough K, Machemer L, Radka S, Jadhav V, Vaish N, Zinnen S, Vargeese C, Bowman K, Shaffer CS, Jeffs LB, Judge A, MacLachlan I, Polisky B. (2005b) Potent and persistent *in vivo* anti-HBV activity of chemically modified siRNAs. Nat Biotechnol 23:1002–1007.

Morton SU, Scherz PJ, Cordes KR, Ivey KN, Stainier DY, Srivastava D. (2008) microRNA-138 modulates cardiac patterning during embryonic development. Proc Natl Acad Sci USA 105:17830–17835.

Murchison EP, Partridge JF, Tam OH, Cheloufi S, Hannon GJ. (2005) Characterization of Dicer-deficient murine embryonic stem cells. Proc Natl Acad Sci USA 102:12135–12140.

Niu Z, Iyer D, Conway SJ, Martin JF, Ivey K, Srivastava D, Nordheim A, Schwartz RJ. (2008) Serum response factor orchestrates nascent sarcomerogenesis and silences the biomineralization gene program in the heart. Proc Natl Acad Sci U S A 105:17824–17829.

Obernosterer G, Leuschner PJ, Alenius M, Martinez J. (2006) Post-transcriptional regulation of microRNA expression. RNA 12:1161–1167.

Owens GK, Kumar MS, Wamhoff BR. (2004) Molecular regulation of vascular smooth muscle cell differentiation in development and disease. Physiol Rev 84:767–801.

Paddison PJ, Caudy AA, Bernstein E, Hannon GJ, Conklin DS. (2002) Short hairpin RNAs (shRNAs) induce sequence-specific silencing in mammalian cells. Genes Dev 16:948–958.

Petrij F, Giles RH, Dauwerse HG, Saris JJ, Hennekam RC, Masuno M, Tommerup N, van Ommen GJ, Goodman RH, Peters DJ, Breuning MH. (1995) Rubinstein-Taybi syndrome caused by mutations in the transcriptional co-activator CBP. Nature 376:348–351.

Prakash TP, Allerson CR, Dande P, Vickers TA, Sioufi N, Jarres R, Baker BF, Swayze EE, Griffey RH, Bhat B. (2005) Positional effect of chemical modifications on short interference RNA activity in mammalian cells. J Med Chem 48:4247–4253.

Rokhlin OW, Scheinker VS, Taghiyev AF, Bumcrot D, Glover RA, Cohen MB. (2008) MicroRNA-34 mediates AR-dependent p53-induced apoptosis in prostate cancer. Cancer Biol Ther 7:1288–1296.

Ross R. (1993) The pathogenesis of atherosclerosis: A perspective for the 1990s. Nature 362:801–809.

Rubinson DA, Dillon CP, Kwiatkowski AV, Sievers C, Yang L, Kopinja J, Rooney DL, Ihrig M.M, McManus M.T, Gertler FB, Scott M.L, Van Parijs L. (2003) A lentivirus-based system to functionally silence genes in primary mammalian cells, stem cells and transgenic mice by RNA interference. Nat Genet 33:401–406.

Saetrom P, Snove O, Nedland M, Grunfeld TB, Lin Y, Bass, MB, Canon JR. (2006) Conserved microRNA characteristics in mammals. Oligonucleotides 16:115–144.

Schmittgen TD, Lee EJ, Jiang J, Sarkar A, Yang L, Elton TS, Chen C. (2008) Real-time PCR quantification of precursor and mature microRNA. Methods 44:31–38.

Schmittgen TD, Jiang J, Liu Q, Yang L. (2004) A high-throughput method to monitor the expression of microRNA precursors. Nucleic Acids Res 32:e43.

Svensson EC, Huggins GS, Lin H, Clendenin C, Jiang F, Tufts R, Dardik FB, Leiden JM. (2000) A syndrome of tricuspid atresia in mice with a targeted mutation of the gene encoding Fog-2. Nat Genet 25:353–356

Siolas D, Lerner C, Burchard J, Ge W, Linsley PS, Paddison PJ, Hannon GJ, Cleary MA. (2005) Synthetic shRNAs as potent RNAi triggers. Nat Biotechnol 23:227–231.

Sluijter JP, van Mil A, van Vliet P, Metz CH, Liu J, Doevendans PA, Goumans MJ. (2010) MicroRNA-1 and -499 Regulate Differentiation and Proliferation in Human-Derived Cardiomyocyte Progenitor Cells. Arterioscler Thromb Vasc Biol 30:859–868.

Snove, O Jr, Rossi JJ (2006) Expressing short hairpin RNAs in vivo. Nat Methods 3:689–695.

Soutschek J, Akinc A, Bramlage B, Charisse K, Constien R, Donoghue M, Elbashir S, Geick A, Hadwiger P, Harborth J, John M, Kesavan V, Lavine G, Pandey RK, Racie T, Rajeev KG, Rohl I, Toudjarska I, Wang G, Wuschko S, Bumcrot D, Koteliansky V, Limmer S, Manoharan M, Vornlocher HP. (2004) Therapeutic silencing of an endogenous gene by systemic administration of modified siRNAs. Nature 432:173–178.

Srivastava S. (2006) Making or breaking the heart: from lineage determination to morphogenesis. Cell 126:1037–1048.

Stegmeier F, Hu G, Rickles RJ, Hannon GJ, Elledge SJ. (2005) A lentiviral microRNA-based system for single-copy polymerase II-regulated RNA interference in mammalian cells. Proc Natl Acad Sci USA 102:13212–13217.

Sun F, Fu H, Liu Q, Tie Y, Zhu J, Xing R, Sun Z, Zheng X. (2008) Downregulation of CCND1 and CDK6 by miR-34a induces cell cycle arrest. FEBS Lett 582:1564–1568.

Takaya T, Ono K, Kawamura T, Takanabe R, Kaichi S, Morimoto T, Wada H, Kita T, Shimatsu A, Hasegawa K. (2009) MicroRNA-1 and MicroRNA-133 in spontaneous myocardial differentiation of mouse embryonic stem cells. Circ J 73:1492–1497

Tatsuguchi M, Seok HY, Callis TE, Thomson JM, Chen JF, Newman M, Rojas M, Hammond SM, Wang DZ. (2007) Expression of microRNAs is dynamically regulated during cardiomyocyte hypertrophy. J Mol Cell Cardiol 42:1137–1141.

Tazawa H, Tsuchiya N, Izumiya M, Nakagama H. (2007) Tumor-suppressive miR-34a induces senescence-like growth arrest through modulation of the E2F pathway in human colon cancer cells. Proc Natl Acad Sci USA 104:15472–15477.

Thomson JM, Newman M, Parker JS, Morin-Kensicki EM, Wright T, Hammond SM. (2006) Extensive post-transcriptional regulation of microRNAs and its implications for cancer. Genes Dev 20:2202–2207.

Thompson RC, Deo M, Turner DL. (2007) Analysis of microRNA expression by in situ hybridization with RNA oligonucleotide probes. Methods 43:153–161.

Thum T, Galuppo P, Wolf C, Fiedler J, Kneitz S, van Laake LW, Doevendans PA, Mummery CL, Borlak J, Haverich A, Gross C, Engelhardt S, Ertl G, Bauersachs J. (2007) MicroRNAs in the human heart: a clue to fetal gene reprogramming in heart failure. Circulation 116:258–267.

Tryndyak VP, Ross SA, Beland FA, Pogribny IP. (2008) Down-regulation of the microRNAs miR-34a, miR-127, and miR-200b in rat liver during hepatocarcinogenesis induced by a methyl-deficient diet. Mol Carcinog 2008 Oct 21. [Epub ahead of print]

Uprichard SL, Boyd B, Althage A, Chisari FV. (2005) Clearance of hepatitis B virus from the liver of transgenic mice by short hairpin RNAs. Proc Natl Acad Sci USA 102:773–778.

Ventura A, Young A, Winslow M, Lintault L, Meissner A, Erkeland S, Newman J, Bronson R, Crowley D, Stone J. (2008) Targeted deletion reveals essential and overlapping functions of the miR-17 through 92 family of miRNA clusters. Cell 132:875–86.

Voorhoeve PM, le Sage C, Schrier M, Gillis AJ, Stoop H, Nagel R, Liu YP, van Duijse J, Drost J, Griekspoor A, Zlotorynski E, Yabuta N, De Vita G, Nojima H, Looijenga LH, Agami R. (2006) A genetic screen implicates miRNA-372 and miRNA-373 as oncogenes in testicular germ cell tumors. Cell 124:1169–1181.

Wang Z, Luo X, Lu Y, Yang B. (2008) miRNAs at the heart of the matter. J Mol Med 86:771–783.

Welch C, Chen Y, Stallings RL. (2007) MicroRNA-34a functions as a potential tumor suppressor by inducing apoptosis in neuroblastoma cells. Oncogene 26:5017–5022.

Wienholds E, Kloosterman WP, Miska E, Alvarez-Saavedra E, Berezikov E, de Bruijn E, Horvitz HR, Kauppinen S, Plasterk RH. (2005) MicroRNA expression in zebrafish embryonic development. Science 309:310–311.

Won Kim H, Haider HK, Jiang S, Ashraf M. (2009) Ischemic preconditioning augments survival of stem cells via miR-210 expression by targeting caspase-8-associated protein 2. J Biol Chem 284:33161–33168.

Wulczyn FG, Smirnova L, Rybak A, Brandt C, Kwidzinski E, Ninnemann O, Strehle M, Seiler A, Schumacher S, Nitsch R. (2007) Post-transcriptional regulation of the let-7 microRNA during neural cell specification. FASEB J 21:415–426.

Xiao J, Lin H, Luo X, Chen G, Wang Z. (2008) miR-605 induces apoptosis of cancer cells by up-regulating p53 through targeting Mdm2 oncogene. EMBO J.

Xia H, Mao Q, Paulson HL, Davidson BL. (2002) siRNA-mediated gene silencing *in vitro* and in vivo. Nat Biotechnol 20:1006–1010.

Xiao J, Luo X, Lin H, Xu C, Gao H, Wang H, Yang B, Wang Z. (2007a) MicroRNA miR-133 represses HERG K$^+$ channel expression contributing to QT prolongation in diabetic hearts. J Biol Chem 282:12363–12367.

Xiao J, Yang B, Lin H, Lu Y, Luo X, Wang Z. (2007b) Novel approaches for gene-specific interference via manipulating actions of microRNAs: examination on the pacemaker channel genes HCN2 and HCN4. J Cell Physiol 212:285–292.

Xu C, Lu Y, Lin H, Xiao J, Wang H, Luo X, Li B, Yang B, Wang Z. (2007) The muscle-specific microRNAs miR-1 and miR-133 produce opposing effects on apoptosis via targeting HSP60/HSP70 and caspase-9 in cardiomyocytes. J Cell Sci 120:3045–3052.

Xu J, Hu Z, Xu Z, Gu H, Yi L, Cao H, Chen J, Tian T, Liang J, Lin Y, Qiu W, Ma H, Shen H, Chen Y. (2009) Functional variant in microRNA-196a2 contributes to the susceptibility of congenital heart disease in a Chinese population. Hum Mutat 30:1231–1236.

Yang B, Lin H, Xiao J, Luo X, Li B, Lu Y, Wang H, Wang Z. (2007) The muscle-specific microRNA miR-1 causes cardiac arrhythmias by targeting GJA1 and KCNJ2 genes. Nat Med 13:486–491.

Yekta S, Shih IH, Bartel DP. (2004) MicroRNA-directed cleavage of HOXB8 mRNA. Science 304:594–596.

Yoshida T, Owens GK. (2005) Molecular determinants of vascular smooth muscle cell diversity. Circ Res 96:280–291.

Zeng Y, Cai X, Cullen BR. (2005). Use of RNA polymerase II to transcribe artificial microRNAs. Methods Enzymol 392:371–380.

Zeng Y, Wagner EJ, Cullen BR. (2002) Both natural and designed micro RNAs can inhibit the expression of cognate mRNAs when expressed in human cells. Mol Cell 9:1327–1333.

Zimmermann TS, Lee AC, Akinc A, Bramlage B, Bumcrot D, Fedoruk MN, Harborth J, Heyes JA, Jeffs LB, John M, Judge AD, Lam K, McClintock K, Nechev LV, Palmer LR, Racie T, Rohl I, Seiffert S, Shanmugam S, Sood V, Soutschek J, Toudjarska I, Wheat AJ, Yaworski E, Zedalis W, Koteliansky V, Manoharan M, Vornlocher HP, MacLachlan I. (2006) RNAi-mediated gene silencing in non-human primates. Nature 441:111–114.

Zhang Y, Lu Y, Wang N, Lin H, Pan Z, Gao X, Zhang F, Zhang Y, Xiao J, Shan H, Luo X, Chen G, Qiao G, Wang Z, Yang B. (2008) Control of experimental atrial fibrillation by microRNA-328. Circulation (accepted).

Zhao Y, Ransom JF, Li A, Vedantham V, von Drehle M, Muth AN, Tsuchihashi T, McManus MT, Schwartz RJ, Srivastava D. (2007) Dysregulation of cardiogenesis, cardiac conduction, and cell cycle in mice lacking miRNA-1-2. Cell 129:303–317.

Zhao Y, Samal E, Srivastava D. (2005) Serum response factor regulates a muscle specific microRNA that targets Hand2 during cardiogenesis. Nature 436:214–220.

Zhou H, Xia XG, Xu Z. (2005) An RNA polymerase II construct synthesizes short-hairpin RNA with a quantitative indicator and mediates highly efficient RNAi. Nucleic Acids Res 33:e62.

CHAPTER 4

miRNAs in Cardiac Hypertrophy and Heart Failure

Abstract: The aim of this chapter is to introduce the role of miRNAs in the pathological process of cardiac hypertrophy/heart failure. Cardiovascular disease is among the main causes of morbidity and mortality in developed countries. In response to stress, the adult heart undergoes remodelling process and hypertrophic growth to adapt to altered workloads and to compensate for the impaired cardiac function. Pathological hypertrophy results in loss of cardiac function and is the major predictor of heart failure and sudden death. Recent studies have established the role of miRNAs in cardiac hypertrophy/heart failure as causal factors or important regulators. miRNAs are aberrantly expressed in various animal models and in patients with heart failure. The miRNAs involved in cardiac hypertrophy/heart failure can in general be divided into two categories: anti-hypertrophic and pro-hypertrophic miRNAs. This chapter introduces the roles of miRNAs in experimental and clinical cardiac hypertrophy/heart failure. Detailed description is given of the well-studied pro-hypertrophic miRNAs miR-195, miR-208 and miR-23a, and anti-hypertrophic miRNAs miR-1, miR-133 and miR-9.

INTRODUCTION

In response to stress (such as hemodynamic alterations associated with myocardial infarction, hypertension, aortic stenosis, valvular dysfunction, etc), the adult heart undergoes remodeling process and hypertrophic growth to adapt to altered workloads and to compensate for the impaired cardiac function. Cardiomyocyte hypertrophy is the dominant cellular response to virtually all forms of hemodynamic overload, endocrine disorders, myocardial injury, or inherited mutations in a variety of structural and contractile proteins. Pathological hypertrophy results in loss of cardiac function and is the major predictor of heart failure and sudden death [McKinsey & Olson, 2005]. Hypertrophic growth manifests enlargement of cardiomyocyte size and enhancement of protein synthesis through the activation of intracellular signaling pathways and transcriptional mediators in cardiac myocytes. The process is characterized by a reprogramming of cardiac gene expression and the activation of 'fetal' cardiac genes [McKinsey & Olson, 2005]. Recent studies revealed an important role for specific miRNAs in the control of hypertrophic growth and chamber remodeling of the heart and point to miRNAs as potential therapeutic targets in heart disease.

Heart failure, i.e., the inability of the heart to pump sufficient blood to the organism, is a frequent and fatal outcome of hypertrophy developed under pathological circumstances. Qualitative alterations—such as reexpression of a fetal gene program—inexorably progress from the hypertrophic stage and characterize this syndrome. Indeed, a transcriptome analysis revealed a similarity in the gene expression of failing and fetal human heart in that 353 mRNAs were found to be more than twofold regulated in common in these two situations with respect to normal adult heart tissue [Thum *et al.*, 2007].

Cardiovascular disease is among the main causes of morbidity and mortality in developed countries. Searching for the causal factors of cardiovascular disease is a restless effort by both scientists and physicians. Recent studies have established the role of miRNAs in cardiac hypertrophy/heart failure as causal factors or important regulators.

miRNAs IN EXPERIMENTAL HYPERTROPHY/HEART FAILURE

Cheng *et al* [2007] identified 19 deregulated miRNAs in hypertrophic mouse hearts after aortic banding. Knockdown of miR-21 expression via AMO-mediated depletion has a significant negative effect on cardiomyocyte hypertrophy induced by TAC in mice or by angiotensin II or phenylephrine in cultured neonatal cardiomyocytes. Consistently, another independent group identified 17 miRNAs up-regulated and 3 miRNAs downregulated in TAC mice, and 7 up-regulated and 4 down-regulated in phenylephrine-induced hypertrophy of neonatal cardiomyocytes. They further showed that inhibition of endogenous miR-21 or miR-18b that are most robustly up-regulated augments hypertrophic growth, while introduction of either of these two miRNAs into cardiomyocytes represses cardiomyocyte hypertrophy [Tatsuguchi *et al.*, 2007].

In a study reported by da Costa Martins *et al* [2008], triggered loss of *Dicer* in the postnatal myocardium using a tamoxifen-inducible Cre recombinase expressed in murine heart cells. Targeted deletion of *Dicer* in 3-week-old mice resulted in spontaneous cardiac remodeling, impairment of cardiac function, and premature death within 1 week. In the adult myocardium, loss of *Dicer* induced rapid and dramatic biventricular enlargement, accompanied by hypertrophic growth of cardiomyocytes, myofiber disarray, ventricular fibrosis, strong induction of fetal gene transcripts, and functional defects. It appears that modifications in miRNA biogenesis affect both juvenile and adult myocardial morphology and function, and miRNAs in general act to restrict expression of growth-stimulating factors.

In both animal models and the human heart, it is generally held that changes in the biology of the cardiac myocyte are the primary initiating events that lead to cardiac remodeling, although it should be noted that cardiac remodeling can occur in the absence of myocyte dysfunction in some experimental models [Urabe *et al.*, 1993; Anand *et al.*, 1997]. One of the principal changes that occurs in the biology of the failing cardiac myocyte is an increase in cell size (hypertrophy). Based on the extant literature, there is evidence that various miRNAs control and/or modulate key components of the hypertrophic process in cardiac myocytes, including reactivation of the so-called fetal gene program. Indeed, the extant experimental literature suggests that miR-1, miR-18b, miR-21, miR-133, miR-195, and miR-208 play important roles in modulating cardiac hypertrophic growth.

Pro-Hypertrophic miRNAs

Role of miR-195

Olson's group presented the first documentation on role of miRNAs in cardiac hypertrophy and in cardiac disease of mammals [van Rooij *et al.*, 2006]. They reported 11 upregulated miRNAs (miR-21, miR-23a, miR-23b, miR-24, miR-27a, miR-27b, miR-125b, miR-195, miR-199a, miR-214, and miR-217) and five downregulated miRNAs (miR-29c, miR-93, miR-133a, miR-150, and miR-181b) in cardiac tissue from mice in response to transverse aortic constriction (TAC) or expression of activated calcineurin, stimuli that induce pathological cardiac remodeling [van Rooij *et al.*, 2006]. Many of these miRNAs were found similarly regulated in failing human hearts. Forced overexpression of stress-inducible miRNAs induced hypertrophy in cultured cardiomyocytes. Particularly, overexpression of miR-195 alone, which is up-regulated during cardiac hypertrophy, is sufficient to induce pathological cardiac growth and heart failure in transgenic mice. myocytes exposed to miRNA overexpression underwent dramatic morphological changes. Forced expression of miR-23a, miR-23b, miR-24, miR-195, and miR-214 induced hypertrophic growth of cardiac cells *in vitro*. Hypertrophy in response to these miRNAs was comparable to that evoked by the adrenergic agonist phenylephrine, among the most potent hypertrophic stimuli known. Overexpression of miR-199a in transgenic mouse line under the control of the α-myosin heavy chain (MHC) promoter resulted in an especially pronounced morphological response in which cardiomyocytes became elongated, reminiscent of a phenotype of eccentric cardiac hypertrophy that results from serial assembly of sarcomeres, associated with dilated cardiomyopathy. Overexpression of miR-125b and miR-133a had no effect on cardiomyocyte morphology. In contrast to the dramatic effects of miR-195 on cardiac structure, function, and gene expression, transgenic cardiac overexpression of miR-214 at levels comparable with those of miR-195 had no phenotypic effect. They were unable to obtain F1 offspring for miR-24 transgene, suggesting that cardiac overexpression of this miRNA causes embryonic lethality.

Cardiac-restricted overexpression of miR-195 resulted in a dilated cardiac hypertrophic phenotype at 6 weeks of age that was accompanied by exaggerated myocyte hypertrophy and activation of the fetal gene ensemble. In contrast, cardiacrestricted overexpression of miR-214 had no effect on cardiac phenotype. Of particular note, miR-195 was significantly upregulated in tow other studies [Lagos-Quintana *et al.*, 2002; Sayed *et al.*, 2007], including human heart failure. This is therefore likely that miR-195 may play an important role in adverse cardiac remodeling.

Role of miR-208

The same group later found that miR-208, encoded by an intron of the alpha myosin heavy chain (αMHC) gene, is required for cardiomyocyte hypertrophy, fibrosis, and expression of αMHC in response to stress and hypothyroidism [van Rooij *et al.*, 2007]. The study showed that miR-208 mutant mice failed to undergo stress-induced cardiac remodeling, hypertrophic growth, and βMHC upregulation, whereas transgenic expression of miR-208 was sufficient to induce βMHC.

Role of miR-23a

Lin *et al* [2009] reported that miR-23a is a pro-hypertrophic miRNA, and its expression is regulated by the transcription factor, nuclear factor of activated T cells (NFATc3), a downstream mediator of calcineurin signaling pathway and a muscle specific ring finger protein 1 as an anti-hypertrophic protein [Molkentin *et al.*, 1998; Delling *et al.*, 2000]. The results showed that miR-23a expression was up-regulated upon treatment with the hypertrophic stimuli including isoproterenol and aldosterone. Knockdown of miR-23a attenuated hypertrophy, suggesting that miR-23a is a prohypertrophic miRNA. They further identified that miR-23a transcription was directly activated by NFATc3. On the other hand, NFATc3 was established to be a target of miR-23a for post-transcription repression. The investigators have also excluded the participation of miR-27a and miR-24 in initiating hypertrophy induced by isoproterenol and aldosterone though their expression is also upregulated [Lin *et al.*, 2009; van Rooij *et al.*, 2006].

Other Pro-Hypertrophic miRNAs

Based on microarray data from aortic banding studies, calcineurin transgenic mouse models and human heart failure samples, van Rooij and colleagues [2006] identified 7 miRNAs that were upregulated and 4 miRNAs that were coordinately downregulated. Five of the miRNAs that were upregulated, miR-23a, -23b, -24, -195, and -214, provoked cardiac myocyte hypertrophy when transfected into neonatal cardiac myocytes, whereas transfection of miR-199a resulted in elongated spindle shapes myocytes that were reminiscent of the elongated cardiac myocytes observed in dilated cardiomyopathy.

Thum *et al* [2007] expressed 3 miRNAs that were upregulated in human heart failure (miR-21, miR-129, and miR-212) and showed that although transfection of a single miRNA had little effect on cell morphology, the simultaneous overexpression of all 3 miRNAs resulted in myocyte hypertrophy and reinduction of the fetal gene program.

Anti-Hypertrophic miRNAs

Role of miR-133

Downregulation of miR-133 and its role in cardiac hypertrophy is probably one of the most significant findings in miRNA research pertinent to cardiovascular disease. The first evidence for the role of miR-133 was documented by Condorelli's group [Carè *et al.*, 2007]. In their study, overexpression of miR-133 resulted in suppression of protein synthesis and inhibition of hypertrophic growth in PE- or endothelin-1–treated neonatal mouse cardiac myocytes and upregulation of fetal genes, including those encoded by ANF, skeletal and cardiac α-actin, and α-myosin heavy chain (αMHC) and βMHC [Carè *et al.*, 2007]. Loss-of-function studies showed that sequestering endogenous miR-133 using a targeted 3'UTR decoy sequence resulted in marked cell hypertrophy and increased protein synthesis, increased fetal gene expression, and perinuclear localization of ANF, consistent with a potential role for miR-133 in suppressing hypertrophic growth [Carè *et al.*, 2007]. Furthermore, treatment of mice using an antisense RNA oligonucleotide (termed an antagomir) targeted to miR-133 resulted in cardiac hypertrophy and reinduction of the fetal gene program. Taken together, these studies suggest that downregulation of miR-1 and miR-133 allows for the increased expression (release) of growthrelated genes that are responsible for cardiac hypertrophy.

However, it is not clear whether these miRNAs contribute to the adverse cardiac remodeling that occurs in heart failure, insofar as the majority of studies in human tissue, save for 1 study,13 have not show that miR-1 and miR-133 are downregulated. Of note, the 1 study that did show that miR-1 and miR-133 were downregulated in human tissue examined dilated atria and myomectomy samples from patients with hypertrophic cardiomyopathy, rather than LV myocardial specimens from patients with heart failure [Carè *et al.*, 2007].

Diabetic cardiomyopathy, characterized by cardiac hypertrophy and contractile dysfunction, eventually leads to heart failure. It has previously been shown that alterations of a number of key molecules are involved in producing cardiomyocyte hypertrophy in diabetes. Feng *et al* [2010] perfomed a study with STZ-induced diabetic mice to establish the development of cardiomyopathy. The tissues were examined for gene expression and miRNA profiling. Neonatal rat cardiomyocytes were used to identify the mechanisms of glucose-induced hypertrophy and the potential role of miR133a. They found that cardiac tissues from the diabetic mice showed alteration of multiple miRNAs by array analysis including miR133a, which was confirmed by RT-PCR. *In vitro* exposure of cardiomyocytes to high levels of glucose produced hypertrophic changes and reduced expression of miRNA133a. Finally, transfection of miR133a mimics prevented altered gene expression and hypertrophic changes. However, no mechanistic data were provided in this study.

Role of miR-1

Abdellatif's group reported an array of miRNAs that are differentially and temporally regulated during cardiac hypertrophy [Sayed *et al.*, 2007]. They found that miR-1 was singularly downregulated as early as day 1, persisting through day 7, after TAC-induced hypertrophy in a mouse model. The functional significance of the findings with respect to miR-1 has been demonstrated in gain-of-function studies *in vitro* and *in vivo*. For example, overexpression of miR-1 in cultured neonatal myocytes partially inhibited phosphorylation of ribosomal S6 protein and inhibited the computationally predicted in silico targets of miR-1 that were related to cardiac growth, including Ras GTPase-activating protein, cyclin-dependent kinase 9, fibronectin, and Ras homolog expressed in brain. miR-1 overexpression also suppressed myocyte spreading induced by serum and endothelin-1 and expression of atrial natriuretic factor (ANF) [Sayed *et al.*, 2007].

Similar downregulation of miR-1 was also reported by the study from Condorelli's group focusing on the role of miR-133 and miR-1 in cardiac hypertrophy with three murine models: TAC mice, transgenic mice with selective cardiac overexpression of a constitutively active mutant of the Akt kinase, and human tissues from patients with cardiac hypertrophy [Carè *et al.*, 2007]. They showed that cardiac hypertrophy in all three models results in reduced expression levels of both miR-133 and miR-1 in the left ventricle. *In vitro* overexpression of miR-133 or miR-1 inhibits cardiac hypertrophy.

Calcium signaling is a central regulator of cardiomyocyte growth and function. Calmodulin is a critical mediator of calcium signals. Calcium, in conjunction with the calcium-binding protein calmodulin (CaM), is a critical mediator of hypertrophy signaling. Ca/CaM activates the phosphatase calcineurin (CN), leading to nuclear accumulation and the activation of the transcription factor nuclear factor of activated T cells (NFAT). The activation of the CN-NFAT pathway is necessary and sufficient for cardiomyocyte hypertrophy [Molkentin *et al.*, 1998; Wilkins *et al.*, 2004]. The transcription factor myocyte enhancer factor 2 (Mef2) is a second major target of Ca/CaM signaling [Passier *et al.*, 2000], and Mef2 is a crucial regulator of cardiomyocyte growth [Xu *et al.*, 2006; Zhang *et al.*, 2002]. Both NFAT and Mef2 regulate cardiac gene expression in collaboration with Gata4, another key regulator of cardiac hypertrophy [Bisping *et al.*, 2006; Liang *et al.*, 2001; Oka *et al.*, 2006]. Because the amount of calmodulin within cardiomyocytes is limiting, the precise control of calmodulin expression is important for the regulation of calcium signaling. In this study, we show for the first time that calmodulin levels are regulated posttranscriptionally in heart failure. The cardiomyocytes restricted miR-1 inhibited the translation of calmodulin-encoding mRNAs via highly conserved target sites within their 3'untranslated regions. In keeping with its effect on calmodulin expression, miR-1 downregulated calcium-calmodulin signaling through calcineurin to NFAT. miR-1 also negatively regulated the expression of Mef2a and Gata4, key transcription factors that mediate calcium-dependent changes in gene expression. Consistent with the downregulation of these hypertrophy-associated genes, miR-1 attenuated cardiomyocyte hypertrophy in cultured neonatal rat cardiomyocytes and in the intact adult heart. Our data indicate that miR-1 regulates cardiomyocyte growth responses by negatively regulating the calcium signaling components calmodulin, Mef2a, and Gata4 [Ikeda *et al.*, 2009].

Role of miR-9

After their study on miR-23a [Lin *et al.*, 2009], the same group subsequently revealed the role of miR-9 in targeting NFATc3 and myocardin in hypertrophic heart with identical experimental approaches [Wang *et al.*, 2010]. They showed that myocardin expression is elevated in response to hypertrophic stimulation with isoproterenol and aldosterone. They then demonstrated that NFATc3 could interact with the promoter region of myocardin and transcriptionally activate its expression. Knockdown of myocardin attenuated hypertrophic responses triggered by NFATc3, suggesting that myocardin is a downstream mediator of NFATc3 in the hypertrophic cascades. Their data further revealed that miR-9 suppressed myocardin translation. The hypertrophic stimulation with isoproterenol and aldosterone decreased the expression levels of miR-9. Administration of miR-9 attenuated cardiac hypertrophy and ameliorated cardiac function. Taken together, the data indicate that miR-9 is an antihypertrophic miRNA acting through repressing myocardin expression [Wang *et al.*, 2010].

Other Anti-Hypertrophic miRNAs

Of the miRNAs that were downregulated in hypertrophic heart, miR-150 and-181b caused a reduction in cardiac myocyte size when transfected into cells [van Rooij *et al.*, 2006].

miRNAs with Controversial Roles in Hypertrophy

One miRNA that is consistently induced by cardiac stress, miR-21, appears to function as a regulator of cardiac growth and fetal gene activation in primary cardiomyocytes in vitro. However, the results on the role of miR-21 in cardiac myocyte hypertrophy have been controversial. Two studies have shown that miR-21 is highly upregulated in mouse models of pressure-overload hypertrophy post-TAB [Sayed *et al.*, 2007; Cheng *et al.*, 2007]. Moreover, angiotensin II– and phenylephrine induced cardiac myocyte hypertrophy is accompanied by a 4~5-fold increase in miR-21 expression in isolated cardiac myocytes. Antisense mediated knockdown of miR-21 partially inhibited phenylephrine- or angiotensin II–induced cell growth and protein synthesis in neonatal rat ventricular myocytes [Cheng *et al.*, 2007].

In contrast, a different group of investigators has shown that antisense knockdown of miR-21 provokes hypertrophy and increased expression of fetal genes in cultured neonatal cardiac myocytes [Tatsuguchi *et al.*, 2007]. The reason(s) for the discrepancy between these studies is unclear but may be that the outcome of miR-21 regulation depends on the animal or cellular models and stages of hypertrophy and may be the modulation of hypertrophic growth by miR-21 in myocytes is through an indirect mechanism, rather than a direct targeting effect of miR-21 on hypertrophy-related genes. Intriguingly in this regard, miR-21 has no validated gene targets relevant to cardiac myocyte hypertrophy. Unfortunately, further investigations into the role of miR-21 in myocyte hypertrophy have been hampered, insofar as exogenously administered pre-miR-21 fails to be processed in cardiac myocytes, resulting in the inability to overexpress mature miR-21 in these cells [van Rooij *et al.*, 2006].

miRNAs IN CLINICAL HYPERTROPHY/HEART FAILURE

Human studies of miRNA expression profiles in heart failure are limited by the lack of standardized protocols, the small numbers of patients in the studies, and the high degree of variability in expression levels between patients.

Earlier in 2007, two independent groups have performed microarrays on RNA isolated from nonfailing and failing hearts. The initial report of miRNA profiling in human heart disease was by Thum *et al* [2007] who compared mRNA and miRNA expression signatures from four non-failing and six failing hearts. The Ambion miRNA microarrays contained probe sets for 384 miRNAs. The authors noted significant (defined as >1.5-fold increase or >50% decrease) upregulation of 67 miRNAs, with downregulation of 43 miRNAs, in the failing vs. control hearts. The miRNA expression signature was similar to that of fetal hearts, suggesting that both mRNA and miR expression in heart failure partially recapitulates that of the embryonic heart. A study directed to the human heart identified 67 significantly upregulated miRNAs and 43 significantly downregulated miRNAs in failing left ventricles versus normal hearts. Interestingly, 86.6% of induced miRNAs and 83.7% of repressed miRNAs are regulated in the same direction in fetal and failing heart tissue compared with healthy hearts, consistent with the activation of 'fetal' cardiac genes in heart failure.

In a substantially larger clinical study published just months later, Ikeda *et al* [2007] described the miRNA profiles of 25 human dilated cardiomyopathy (DCM), 19 ischaemic cardiomyopathy (ICM), and 13 pressure overload hypertrophy (AS), compared with that of 10 normal hearts. Their assay measured levels of 428 individual miRNAs using a high-throughput bead-based platform [Lu *et al.*, 2005] that detected 87 cardiac-expressed miRNAs, 43 of which were regulated in at least one of the disease groups. Importantly, the miRNA expression profile appeared to be distinguishable between disease groups, and within the primary data set the miRNA signature was able to predict the diagnosis with an accuracy rate that approached 70%. Among the miRNAs, miR-214 was found increased >2 folds in all three conditions versus control. miR-1 was downregulated, whereas miR-125b was upregulated, by 30-40% in DCM and AS, but not in ICM. No changes of miR-133a/b were observed in any of these pathological conditions.

Subsequently in 2008, Sucharov *et al* [2008] compared miRNA expression in five ischaemic and five non-ischaemic cardiomyopathic hearts, compared with six non-failing hearts. The assay used a microarray containing probes for

470 miRNAs. Thirty-three miRs were reported as regulated in either ischaemic and/or nonischaemic cardiomyopathic hearts, several of which were shown to have measurable effects in cultured neonatal rat ventricular cardiomyocytes. Naga Prasad *et al* [2009]. used a custom microarray to identify eight miRs (seven of which had been previously identified in human or mouse heart failure) that were upregulated in 50 heart failure cardiac samples, compared with twenty non-failing specimens, and independently validated the associations in 20 dilated cardiomyopathy and 10 non-failing samples. Taken together, these studies support the idea that miR regulation may be sufficiently distinct in different forms of cardiac injury to be able to discriminate between heart failure of ischaemic vs. nonischaemic aetiology.

Later in 2009, Divakaran and Mann [2009] analyzed miRNA expression profiles on 172 miRNAs from a total of 10 different experimental and clinical models published in the literature, of which 5 were in human tissue, 4 were *in vivo* studies in animals, and 1 was an *in vitro* study in cultured cardiac myocytes. Their analysis provides a succinct and resourceful summary of miRomes in various cardiac conditions in human and mouse. The analysis reveals that for miRNAs with increased levels of expression levels, there was good concordance between human heart failure and experimental models of pathological remodeling. Indeed, 25 miRNAs (7b, 7c, 10b, 15b, 21, 23a, 23b, 24, 27a, 27b, 29a, 103, 125b, 140*, 195, 199a, 199a*, 199b, 208, 210, 211, 214, 330, 341, 424) were upregulated in one or more myocardial samples from failing human hearts and experimental models, suggesting that changes in miRNA expression patterns in experimental models may provide further insight into our understanding LV remodeling in human heart failure. Interestingly, the expression profile of miRNAs with decreased expression levels were far less concordant in experimental models and human heart failure samples. Indeed, there were only 10 miRNA species (1, 10a, 26b, 30a_5p, 30b, 30c,150, 218, 451, 499) that were downregulated in one or more myocardial samples from failing human hearts and experimental models. It should be emphasized that different microarray platforms were used in the above studies, which may account for some of the observed differences in miRNA regulation.

Apart from these alterations in miRNA expression, a decreased level of Dicer is a newly reported feature of heart failure in patients with dilated cardiomyopathy [Chen *et al.*, 2008]. Dicer is the only known enzyme involved in the maturation of miRNAs from their precursors, so reduced Dicer expression reduces the expression of all mature miRNAs as shown by microarray data [Chen *et al.*, 2008]. However, how a global decrease in miRNA maturation can be reconciled with increased expression of those miRNAs found to be upregulated with disease needs to be addressed. One explanation might be that decreased Dicer expression occurs as progression of disease nears failure and is, therefore, a late phenomenon. In support of this, Dicer-knockout mice pups, which presented with dilated hearts, died rapidly by postnatal day 4; hearts from patients presenting with aortic stenosis, which presumably were hypertrophic, had higher levels of Dicer with respect to failing hearts; and Dicer reverted to significantly higher levels in failing hearts after a left ventricular assist device was installed for mechanical support in anticipation of transplantation [Chen *et al.*, 2008]. This might be an important finding that could shed light on why the heart starts failing when it does. In addition, the miRNome has been reported to undergo temporal changes during the development of experimentally induced hypertrophy in mice [Cheng *et al.*, 2007; Sayed *et al.*, 2007]; it cannot be ruled out, therefore, that there may also be some changes in the miRNome that predispose to heart failure progression.

Collectively, with respect to hypertrophy, it is evident that in addition to the muscle-specific miRNAs miR-1, miR-133 and miR-208, other miRNAs, including miR-195, miR-21, miR-18b, also play an important role. It appears that multiple miRNAs are involved in cardiac hypertrophy and each of them can independently determine the pathological process. The first common finding is that an array of miRNAs is significantly altered in their expression, either up- or down-regulated and studies from different research groups demonstrated overlapping miRNAs that are altered in cardiac hypertrophy. The second common finding is that single miRNAs can critically determine the progression of cardiac hypertrophy.

It should be noted that controversies exist among these studies. Probably the most striking disparity is the contradictory results from the studies reported by Cheng *et al* [2007] and Tatsuguchi *et al* [2007]. Cheng *et al* showed that knockdown of *miR-21* relieves cardiomyocyte hypertrophy, whereas the study from Wang's group demonstrated the opposite: inhibition of endogenous *miR-21* augments and introduction of functional *miR-21* into cardiomyocytes represses myocyte hypertrophy. Both studies used the phenylephrine model of hypertrophy in cultured neonatal rat ventricular myocytes. The study by Carè *et al* [2007] clearly indicates that *miR-133* is an anti-hypertrophic factor and downregulation of *miR-133* alone is sufficient to induce cardiac hypertrophy. However, the study reported by van Rooij *et al* [2006] suggests that *miR-133* does not cause any morphological changes of cardiomyocytes indicative of hypertrophic growth. Moreover, among the 7 studies, two reported downregulation of *miR-133* in hypertrophy, three failed to observe this change, one found it up-regulated, and another study did not deal with *miR-133*. While *miR-1* was found to be down-regulated in 2 studies, this observation was not reproduced in 3 other studies and one study performed with human heart actually claimed it to be up-regulated. Increase in *miR-18b* was seen only in one study. The most consistent changes reported by the 6 studies are up-regulation of *miR-21* (6 of 6 studies), *miR-23a* (4 of 6), *miR-125b* (5 of 6) and *miR-214* (4 of 6), and down-regulation of *miR-150* (5 of 6 studies) and *miR-30* (5 of 6). The discrepancies could be explained by multiple factors, including different models of hypertrophy, different time points of the same models, different miRNA microarrays for analyses, etc.

REFERENCES

Anand IS, Liu D, Chugh SS, Prahash AJ, Gupta S, John R, Popescu F, Chandrashekhar Y. (1997) Isolated myocyte contractile function is normal in postinfarct remodeled rat heart with systolic dysfunction. Circulation 96:3974–3984.

Bisping E, Ikeda S, Kong SW, Tarnavski O, Bodyak N, McMullen JR, Rajagopal S, Son JK, Ma Q, Springer Z, Kang PM, Izumo S, Pu WT. (2006) Gata4 is required for maintenance of postnatal cardiac function and protection from pressure overload-induced heart failure. Proc Natl Acad Sci USA 103:14471–14476.

Brennecke J, Stark A, Russell RB, Cohen SM. (2005) Principles of microRNA-target recognition. PLoS Biol 3:404–418.

Brower GL, Gardner JD, Forman MF, Murray DB, Voloshenyuk T, Levick SP, Janicki JS. (2006) The relationship between myocardial extracellular matrix remodelling and ventricular function. Eur J Cardiothorac Surg 30:604–610.

Bruggink AH, van Oosterhout MF, de Jonge N, Cleutjens JP, van Wichen DF, van Kuik J, Tilanus MG, Gmelig-Meyling FH, van den Tweel JG, de Weger RA. (2007) Type IV collagen degradation in the myocardial basement membrane after unloading of the failing heart by left ventricular assist device. Lab Invest 2007;87:1125–1137.

Bruggink AH, van Oosterhout MF, de Jonge N, Ivangh B, van Kuik J, Voorbij RH, Cleutjens JP, Gmelig-Meyling FH, de Weger RA. (2006) Reverse modeling of the myocardial extracellular matrix after prolonged left ventricular assist device support follows a biphasic pattern. J Heart Lung Transplant 25:1091–1098.

Calin GA, Cimmino A, Fabbri M, Ferracin M, Wojcik SE, Shimizu M, Taccioli C, Zanesi N, Garzon R, Aqeilan RI, Alder H, Volinia S, Rassenti L, Liu X, Liu CG, Kipps TJ, Negrini M, Croce CM. (2008) MiR-15a and miR-16–1 cluster functions in human leukemia. Proc Natl Acad Sci USA 105:5166–5171.

Carè A, Catalucci D, Felicetti F, Bonci D, Addario A, Gallo P, Bang ML, Segnalini P, Gu Y, Dalton ND, Elia L, Latronico MV, Høydal M, Autore C, Russo MA, Dorn GW, Ellingsen O, Ruiz-Lozano P, Peterson KL, Croce CM, Peschle C, Condorelli G. (2007) MicroRNA-133 controls cardiac hypertrophy. Nat Med 13:613–618.

Chen JF, Murchison EP, Tang R, Callis TE, Tatsuguchi M, Deng Z, Rojas M, Hammond SM, Schneider MD, Selzman CH, Meissner G, Patterson C, Hannon GJ, Wang DZ. (2008) Targeted deletion of Dicer in the heart leads to dilated cardiomyopathy and heart failure. Proc Natl Acad Sci USA 105:2111–2116.

Cheng Y, Ji R, Yue J, Yang J, Liu X, Chen H, Dean DB, Zhang C. (2007) MicroRNAs are aberrantly expressed in hypertrophic heart. Do they play a role in cardiac hypertrophy? Am J Pathol 170:1831–1840.

Cimmino A, Calin GA, Fabbri M, Iorio MV, Ferracin M, Shimizu M, Wojcik SE, Aqeilan RI, Zupo S, Dono M, Rassenti L, Alder H, Volinia S, Liu CG, Kipps TJ, Negrini M, Croce CM. (2005) miR-15 and miR-16 induce apoptosis by targeting BCL2. Proc Natl Acad Sci USA 102:13944–13949.

da Costa Martins PA, Bourajjaj M, Gladka M, Kortland M, van Oort RJ, Pinto YM, Molkentin JD, De Windt LJ. (2008) Conditional dicer gene deletion in the postnatal myocardium provokes spontaneous cardiac remodeling. Circulation 118:1567–1576.

de Jonge N, Kirkels H, Lahpor JR, Klöpping C, Hulzebos EJ, de la Rivière AB, Robles de Medina EO. (2001) Exercise performance in patients with end-stage heart failure after implantation of a left ventricular assist device and after heart transplantation: an outlook for permanent assisting? J Am Coll Cardiol 37:1794–1799.

de Jonge N, van Wichen DF, van Kuik J, Kirkels H, Lahpor JR, Gmelig-Meyling FH, van den Tweel JG, de Weger RA. (2003) Cardiomyocyte death in patients with end-stage heart failure before and after support with a left ventricular assist device: low incidence of apoptosis despite ubiquitous mediators. J Heart Lung Transplant 22:1028–1036.

de Jonge N, van Wichen DF, Schipper ME, Lahpor JR, Gmelig-Meyling FH, Robles de Medina EO, de Weger RA. (2002) Left ventricular assist device in end-stage heart failure: persistence of structural myocyte damage after unloading. An immunohistochemical analysis of contractile myofilaments. J Am Coll Cardiol 39:963–969.

Delling U, Tureckova J, Lim HW, De Windt LJ, Rotwein P, Molkentin JD. (2000) A calcineurin-NFATc3-dependent pathway regulates skeletal muscle differentiation and slow myosin heavy-chain expression. Mol Cell Biol 20:6600–6611.

Divakaran V, Mann DL. (2008) The emerging role of microRNAs in cardiac remodeling and heart failure. Circ Res 103:1072–1083.

Doench JG, Petersen CP, Sharp PA. (2003) siRNAs can function as miRNAs. Genes Dev 17:438–442.

Doench JG, Sharp PA. (2004) Specificity of microRNA target selection in translational repression. Genes Dev 18:504–511.

Elbashir SM, Lendeckel W, Tuschl T. (2001) RNA interference is mediated by 21- and 22-nucleotide RNAs. Genes Dev 15:188–200.

Feng B, Chen S, George B, Feng Q, Chakrabarti S. (2010) miR133a regulates cardiomyocyte hypertrophy in diabetes. Diabetes Metab Res Rev 26:40–49.

Grishok A, Pasquinelli AE, Conte D, Li N, Parrish S, Ha I, Baillie DL, Fire A, Ruvkun G, Mello CC. (2001) Genes and mechanisms related to RNA interference regulate expression of the small temporal RNAs that control C. elegans developmental timing. Cell 106:23–34.

Ha I, Wightman B, Ruvkun G. (1996) A bulged lin-4/lin-14 RNA duplex is sufficient for Caenorhabditis elegans lin-14 temporal gradient formation. Genes Dev 10:3041–3050.

Haley B, Zamore PD. (2004) Kinetic analysis of the RNAi enzyme complex. Nat Struct Mol Biol 11:599–606.

Hutvágner G, Zamore PD. (2002) A microRNA in a multiple-turnover RNAi enzyme complex. Science 297:2056–2060.

Ikeda S, He A, Kong SW, Lu J, Bejar R, Bodyak N, Lee KH, Ma Q, Kang PM, Golub TR, Pu WT. (2009) MicroRNA-1 negatively regulates expression of the hypertrophy-associated calmodulin and Mef2a genes. Mol Cell Biol 29:2193–2204.

Ikeda S, Kong SW, Lu J, Bisping E, Zhang H, Allen PD, Golub RD, Pieske B, Pu WT. (2007) Altered microRNA expression in human heart disease. Physiol Genomics 31:367–373.

Kapoun AM, Liang F, O'Young G, Damm DL, Quon D, White RT, Munson K, Lam A, Schreiner GF, Protter AA. (2004) B-type natriuretic peptide exerts broad functional opposition to transforming growth factorbeta in primary human cardiac fibroblasts: fibrosis, myofibroblast conversion, proliferation and inflammation. Circ Res 94:453–461.

Kiriakidou M, Tan GS, Lamprinaki S, De Planell-Saguer M, Nelson PT, Mourelatos Z. (2007) An mRNA m(7)G cap binding-like motif within human Ago2 represses translation. Cell 129:1141–1151.

Kloosterman WP, Wienholds E, Ketting RF, Plasterk RH. (2004) Substrate requirements for let-7 function in the developing zebrafish embryo. Nucleic Acids Res 32:6284–6291.

Lagos-Quintana M, Rauhut R, Yalcin A, Meyer J, Lendeckel W, Tuschl T. (2002) Identification of tissue-specific microRNAs from mouse. Curr Biol 12:735–739.

Latif N, Yacoub MH, George R, Barton PJ, Birks EJ. (2007) Changes in sarcomeric and non-sarcomeric cytoskeletal proteins and focal adhesion molecules during clinical myocardial recovery after left ventricular assist device support. J Heart Lung Transplant 26:230–5.

Lewis BP, Shih IH, Jones-Rhoades MW, Bartel DP, Burge CB. (2003) Prediction of mammalian microRNA targets. Cell 115:787–798.

Liang Q, De Windt LJ, Witt SA, Kimball TR, Markham BE, Molkentin JD. (2001) The transcription factors GATA4 and GATA6 regulate cardiomyocyte hypertrophy *in vitro* and in vivo. J Biol Chem 276:30245–30253.

Lin Z, Murtaza I, Wang K, Jiao J, Gao J, Li PF. (2009) miR-23a functions downstream of NFATc3 to regulate cardiac hypertrophy. Proc Natl Acad Sci USA 106:12103–108.

Lu J, Getz G, Miska EA, Alvarez-Saavedra E, Lamb J, Peck D, Sweet-Cordero A, Ebert BL, Mak RH, Ferrando AA, Downing JR, Jacks T, Hovitz HR, Golub TR. (2005) MicroRNA expression profiles classify human cancers. Nature 435:834–838.

Martin SE, Caplen NJ. (2006) Mismatched siRNAs downregulate mRNAs as a function of target site location. FEBS Lett 580:3694–3698.

Martinez J, Tuschl T. (2004) RISC is a 5' phosphomonoester-producing RNA endonuclease. Genes Dev 18:975–980.

Matkovich SJ, Van Booven DJ, Youker KA, Torre-Amione G, Diwan A, Eschenbacher WH, Dorn LE, Watson MA, Margulies KB, Dorn GW. (2009) Reciprocal regulation of myocardial microRNAs and messenger RNA in human cardiomyopathy and reversal of the microRNA signature by biomechanical support. Circulation 119:1263–1271.

McKinsey TA, Olson EN. (2005) Toward transcriptional therapies for the failing heart: chemical screens to modulate genes. J Clin Invest 115:538–546.

Molkentin JD, Lu JR, Antos CL, Markham B, Richardson J, Robbins J, Grant SR, Olson EN. (1998) A calcineurin-dependent transcriptional pathway for cardiac hypertrophy. Cell 93:215–228.

Naga Prasad SV, Duan ZH, Gupta MK, Surampudi VS, Volinia S, Calin GA, Liu CG, Kotwal A, Moravec CS, Starling RC, Perez DM, Sen S, Wu Q, Plow EF, Croce CM, Karnik S. (2009) Unique microRNA profile in end-stage heart failure indicates alterations in specific cardiovascular signaling networks. J Biol Chem 284:27487–27499.

Oka T, Maillet M, Watt AJ, Schwartz RJ, Aronow BJ, Duncan SA, Molkentin JD. (2006) Cardiac-specific deletion of Gata4 reveals its requirement for hypertrophy, compensation, and myocyte viability. Circ Res 98:837–845.

Passier R, Zeng H, Frey N, Naya FJ, Nicol RL, McKinsey TA, Overbeek P, Richardson JA, Grant SR, Olson EN. (2000) CaM kinase signaling induces cardiac hypertrophy and activates the MEF2 transcription factor in vivo. J Clin Invest 105:1395–1406.

Sayed D, Hong C, Chen IY, Lypowy J, Abdellatif M. (2007) MicroRNAs play an essential role in the development of cardiac hypertrophy. Circ Res 100:416–424.

Scarabelli TM, Knight R, Stephanou A, Townsend P, Chen-Scarabelli C, Lawrence K, Gottlieb R, Latchman D, Narula J. (2006) Clinical implications of apoptosis in ischemic myocardium. Curr Probl Cardiol 31:181–264.

Schipper ME, van Kuik J, de Jonge N, Dullens HF, de Weger RA. (2008) Changes in regulatory microRNA expression in myocardium of heart failure patients on left ventricular assist device support. J Heart Lung Transplant 27:1282–1285.

Schnee PM, Shah N, Bergheim M, Poindexter BJ, Buja LM, Gemmato C, Radovancevic B, Letsou GV, Frazier OH, Bick RJ. (2008) Location and density of alpha- and beta-adrenoreceptor sub-types in myocardium after mechanical left ventricular unloading. J Heart Lung Transplant 27:710–717.

Sucharov C, Bristow MR, Port JD. (2008) miRNA expression in the failing human heart: functional correlates. J Mol Cell Cardiol 45:185–192.

Tabara H, Sarkissian M, Kelly WG, Fleenor J, Grishok A, Timmons L, Fire A, Mello CC. (1999) The rde-1 gene, RNA interference, and transposon silencing in C. elegans. Cell 99:123–32.

Tatsuguchi M, Seok HY, Callis TE, Thomson JM, Chen JF, Newman M, Rojas M, Hammond SM, Wang DZ. (2007) Expression of microRNAs is dynamically regulated during cardiomyocyte hypertrophy. J Mol Cell Cardiol 42:1137–1141.

Thum T, Galuppo P, Wolf C, Fiedler J, Kneitz S, vanLaake LW, Doevendans PA, Mummery CL, Borlak J, Haverich A, Gross C, Engelhardt S, Ertl G, Bauersachs J. (2007) MicroRNAs in the human heart: a clue to fetal gene reprogramming in heart failure. Circulation 116:258–267.

Urabe Y, Hamada Y, Spinale FG, Carabello BA, Kent RL, Cooper G IV, Mann DL. (1993) Cardiocyte contractile performance in experimental biventricular volume-overload hypertrophy. Am J Physiol 264:H1615–H1623.

van Rooij E, Sutherland LB, Liu N, Williams AH, McAnally J, Gerard RD, Richardson JA, Olson EN. (2006) A signature pattern of stress-responsive microRNAs that can evoke cardiac hypertrophy and heart failure. Proc Natl Acad· Sci USA 103:18255–18260.

van Rooij E, Sutherland LB, Qi X, Richardson JA, Hill J, Olson EN. (2007) Control of stress-dependent cardiac growth and gene expression by a microRNA. Science 316:575–579.

Wang K, Long B, Zhou J, Li PF. (2010) miR-9 and NFATc3 regulate myocardin in cardiac hypertrophy. J Biol Chem 285:11903–11912.

Wilkins BJ, Dai YS, Bueno OF, Parsons SA, Xu J, Plank DM, Jones F, Kimball TR, Molkentin JD. (2004) Calcineurin/NFAT coupling participates in pathological, but not physiological, cardiac hypertrophy. Circ Res 94:110–118.

Xiao J, Lin H, Luo X, Chen G, Wang Z. (2008) New microRNAs in the p53 tumor suppressor network: regulators of p53-mdm2 negative feedback loop. EMBO J (accepted).

Xiao J, Yang B, Lin H, Lu Y, Luo X, Wang Z. (2007) Novel approaches for gene-specific interference via manipulating actions of microRNAs: examination on the pacemaker channel genes HCN2 and HCN4. J Cell Physiol 212:285–292.

Xu J, Gong NL, Bodi I, Aronow BJ, Backx PH, Molkentin JD. (2006) Myocyte enhancer factors 2A and 2C induce dilated cardiomyopathy in transgenic mice. J Biol Chem 281:9152–9162.

Zeng Y, Yi R, Cullen BR. (2003) MicroRNAs and small interfering RNAs can inhibit mRNA expression by similar mechanisms. Proc Natl Acad Sci USA 100:9779–9784.

Zhang CL, McKinsey TA, Chang S, Antos CL, Hill JA, Olson EN. (2002) Class II histone deacetylases act as signal-responsive repressors of cardiac hypertrophy. Cell 110:479–488.

Zhao Y, Samal E, Srivastava D. (2005) Serum response factor regulates a muscle-specific microRNA that targets Hand2 during cardiogenesis. Nature 436:214–220.

CHAPTER 5

miRNAs in Myocardial Infarction

Abstract: The aim of this chapter is to provide an overview on the role of miRNAs in myocardial ischemia, ischemia/reperfusion injury and ischemic preconditioning. Myocardial ischemia due to occlusion of coronary arteries constitutes the major cause of mortality and morbidity of humans worldwide by causing an array of injuries. Timely myocardial reperfusion remains the most effective treatment strategy for reducing myocardial infarct size, preventing left ventricular remodelling, preserving left ventricular systolic function and improving clinical outcomes. However, the full benefits of myocardial reperfusion are not realized, given that the actual process of reperfusing ischemic myocardium can independently induce myocardial injury. On the other hand, heart has endogenous cardioprotective capability against myocardial/reperfusion injury, called ischemic preconditioning. Recent studies indicate that miRNAs are implicated in all these different aspects of myocardial ischemia. This chapter describes the role of miR-1 and mR-133 in myocardial ischemia, miR-21, miR-29 and miR-320 in ischemia/reperfusion injury, and miR-21 and miR-199a in preconditioning.

INTRODUCTION

Although a substantial reduction in death rate from cardiovascular causes during the past 50 years, ischemic heart disease (IHD) remains the leading cause of morbidity and mortality in the industrialized world. In the United States, IHD afflicts in excess of 6 million Americans annually (American Heart Association, 2004), it causes about 152,000 deaths per year in UK, and world-wide, one in eight deaths is attributed to IHD [Ghuran & Camm, 2001]. Myocardial ischemia leads to a cascade of metabolic events which are interrelated and are caused by hypoxia, acidosis, oxidative stress, calcium overload, decreases in survival signaling molecules and increases in death signaling mediators, etc. The ischemic injuries are typically contractile dysfunction, electrical disturbance, and cell loss. Timely coronary reperfusion as treatment for acute myocardial infarction reduces myocardial infarct size, improves left ventricular function and survival. Although reperfusion is an absolute prerequisite for the survival of ischemic tissue, some component of reperfusion may be detrimental and able to inflict injury over and above that attributable to the ischemia. Theoretically, if this "reperfusion injury" could be treated and eliminated, the outcome for patients with myocardial infarction might further improve. The concept of reperfusion injury is closely tied to the concept that oxygen radicals generated at the time of reperfusion cause tissue damage. There are four basic forms of reperfusion injury. Lethal reperfusion injury is described as myocyte cell death due to reperfusion itself rather than to the preceding ischemia. This concept continues to be controversial in both experimental animal and clinical studies. Vascular reperfusion injury refers to progressive damage to the vasculature over time during the phase of reperfusion. Manifestations of vascular reperfusion injury include an expanding zone of no reflow and a deterioration of coronary flow reserve. This form of reperfusion injury has been documented in animal models and probably occurs in humans. Stunned myocardium refers to postischemic ventricular dysfunction of viable myocytes and probably represents a form of "functional reperfusion injury." This phenomenon is well documented in both animal models and humans. Reperfusion arrhythmias represent the fourth form of reperfusion injury. They include ventricular tachycardia and fibrillation that occur within seconds to minutes of restoration of coronary flow after brief (5 to 15 min) episodes of myocardial ischemia [Kloner, 1993].

On the other hand, the heart possess an intrinsic protective mechanism. The concept of cardioprotection was first conceived in the late 1960s and has evolved to include the endogenous cardioprotective phenomenon of ischemic conditioning, a concept in which the heart can be protected from an episode of acute lethal ischemia-reperfusion injury by applying brief non-lethal episodes of ischemia and reperfusion either to the heart itself or to an organ or tissue that is remote from the heart. The brief conditioning episodes of ischemia and reperfusion can be applied prior to the index ischemic episode (ischemic preconditioning), after the onset of the index ischemic episode (ischemic perconditioning), or at the onset of reperfusion (ischemic postconditioning).

Since the first evidence for the role of miR-1 in arrhythmogenesis in AMI [Yang *et al.*, 2007], there have been several studies on miRNAs in myocardial ischemia, ischemia/reperfusion injury and ischemic preconditioning. Some miRNAs were found to cause detrimental effects and some beneficial effects.

miRNAs IN MYOCARDIAL ISCHEMIA

miR-1

In 2007, my laboratory presented the first evidence for miRNA involvement in myocardial ischemic injury by demonstrating the upregulation of miR-1 in the heart of patients with coronary artery disease and in a rat model of acute myocardial infarction (AMI) and the arrhythmogenic property of this deregulation [Yang et a., 2007]. A more recent report showed ~3.8-fold miR-1 up-regulation in the samples of infarcted tissue and remote myocardium from twenty-four patients with AMI [Boštjančič *et al.*, 2010]. However, the same group claimed a downregulation of miR-1 in the autopsy samples of infarcted heart tissue from 50 patients with MI, compared to healthy adult and fetal hearts, in their earlier study [Boštjančič *et al.*, 2009]. Intriguingly, upregulation of miR-1 was subsequently confirmed by other studies in animal models of ischemia/reperfusion injury (I/R-I) [Tang *et al.*, 2009; Yin *et al.*, 2008]. In addition, upregulation of miR-1 in MI is also consistent with recent discovery of circulating miR-1 as a potential biomarker for MI (see **Chapter 15**) [Ai *et al.*, 2010; Wang *et al.*, 2010]. In addition to its arrhythmogenic potential, we have also discovered the proapoptotic property of miR-1 in cardiomyocytes in response to oxidative stress [Xu *et al.*, 2007], another sort of ischemic myocardial injury.

Based on the available data in the literature, it can be speculated that miR-1 likely produces injuries to myocardium as it has been shown to cause arrhythmias in AMI [Yang *et al.*, 2007] and apoptosis of cardiomyocytes in response to oxidative stress [Xu *et al.*, 2007].

miR-133

In the study by Boštjančič *et al* [2010], the authors observed down-regulation of miR-133a/b in infarcted tissue and remote myocardium, indicating miR-133a/b involvement in the heart response to AMI. Since miR-133 has been shown by us to produce antiapoptotic action in response to oxidative stress [Xu *et al.*, 2007], it is anticipated that downregulation of miR-133 in AMI cpuld exaggerate ischemic injury.

miRNAs IN ISCHEMIA/REPERFUSION INJURY (I/R-I)

miR-21

Roy *et al* [2009] performed miRNA expression profiling and quantification using a Bioarray system that screens for human-, mice-, rat- and Ambi- miRNAs in myocardial tissues collected following 2h and 7h of IR or sham operation. Data mining and differential analyses resulted in 14 miRNAs that were upregulated on day 2, 9 miRNAs that were upregulated on day 7 and 7 miRNAs that were down-regulated on day 7 post-IR. Results randomly selected from expression profiling were validated using real-time PCR. Tissue elements laser captured from the infarct site showed marked induction of miR-21. In situ hybridization studies using locked nucleic acid miR-21-specific probe identified that IR-inducible miR-21 was specifically localized in the infarct region of the IR heart. Immunohistochemistry data show that cardiac fibroblasts are the major cell type in the infarct zone. It appears that miR-21 primarily produces cardioprotecive effect via reducing apoptotic cell death (see Chapter 13).

miR-29

Pioglitazone, a peroxisome proliferator-activated receptor (PPAR)-γ agonist, has been documented by numerous studies to be able to limit myocardial infarct size in experimental animals [Yasuda *et al.*, 2009; Wynne *et al.*, 2005]. Ye *et al* [2010] assessed the effects of PPAR-γ activation on myocardial miRNAs levels and the role of miRNAs in I/R-I in the rat heart after pioglitazone administration using miRNA arrays, followed by Northern Blot verification. They found that miR-29a and miR-29c levels were decreased after 7-day treatment with pioglitazone. In H9c2 cells, the effects of pioglitazone and rosiglitazone on miR-29 expression levels were blocked by a selective PPAR-γ inhibitor GW9662. Down-regulation of miR-29 by antisense inhibitor or by pioglitazone protected H9c2 cells from simulated IR injury, with increased cell survival and decreased caspase-3 activity. In contrast, overexpressing miR-29 promoted apoptosis and completely blocked the protective effect of pioglitazone. Antagomirs against miR-29a or miR-29c significantly reduced myocardial infarct size and apoptosis in hearts subjected to I/R-I. Western blot analyses demonstrated that Mcl-2, an anti-apoptotic Bcl-2 family member, was increased by miR-29 inhibition, similar to the finding in cancer cells [Mott *et al.*, 2007]. Down-regulation of miR-29 protected hearts against I/R-I.

miR-320

In contrast to miR-21 and miR-29, miR-320 seems to cause damaging effects to the heart subjected to I/R-I. Ren *et al* [2009] analyzed miRNA expression profile in murine hearts subjected to ischemia/reperfusion (I/R) *in vivo* and *ex vivo*, followed by qRT-PCR verification. They found that only miR-320 expression was significantly decreased in the hearts on I/R-I *in vivo* and *ex vivo*. Overexpression of miR-320 in cultured adult rat cardiomyocytes enhanced apoptotic cell death, whereas knockdown produced cytoprotective effect against apoptosis, on simulated I/R-I. Furthermore, transgenic mice with cardiac-specific overexpression of miR-320 revealed an increased extent of apoptosis and infarction size in the hearts on I/R-I *in vivo* and *ex vivo* relative to the wild-type controls. Conversely, *in vivo* treatment with antagomir-320 reduced infarction size relative to the administration of mutant antagomir-320 and saline controls. They subsequently identified heat-shock protein 20 (Hsp20), a cardioprotective protein, as a target for miR-320, utilizing a luciferase/GFP reporter activity assay and examining the expression of Hsp20 on miR-320 overexpression and knockdown in cardiomyocytes.

miRNAs IN ISCHEMIC PRECONDITIONING (IPC)

Ischemic preconditioning (IPC) is a powerful intinsc cardioprotective mechanism whereby repeated brief episodes of ischemia protect the heart against ssequent myocardial infarction [Murry *et al.*, 1986]. Genetic reprogramming emerging during or following IPC, which simulates angina in the clinical setting can be characterized as protective in nature. Several mechanisms for IPC have been proposed, which broadly include the release of endogenous mediators including adenosine, activation of G-coupled receptors and protein kinase C, and synthesis of cytoprotective proteins including endothelial nitric oxide synthase (eNOS), inducible nitric oxide synthase (iNOS), cyclooxygenase-2, and heat shock protein (HSP)70, which either individually or in concert lead to protection against ischemia/reperfusion (I/R) injury [Bolli, 2000; Heusch *et al.*, 2008].

miR-21

Yin *et al* [2009] recently reported an interesting study in which they first created IPC in a mouse model of Langendorff isolated perfused heart, then extracted miRNAs from the heart. To determine the cause–effect relationship between IPC-induced endogenous miRNAs and cardioprotection, they injected the pool of extracted miRNAs from non-IPC and IPC hearts directly into the left ventricular wall *in situ* in a separate set of mice (miRNA-injected group). And 48 hours later, the hearts were subjected to regional ischemia/reperfusion injury by left anterior descending artery ligation for 30 minutes followed by reperfusion for 24 hour. Prior to injection, the extrcated miRNA sample was treated with polyamine at 22°C for 30 minutes to form miRNA-amine complexes which facilitate miRNAs' entry into cells. miRNAs derived from IPC hearts produced a protective phenotype against ischemia/reperfusion injury with significantly lower infarction size as compared to saline-injected controls or miRNAs prepared from non-IPC hearts. Moreover, IPC caused significant increases in miR-1, miR-21 and miR-24 and the "IPC-miRNA" treatment caused an increase in eNOS mRNA and protein, HSF-1 (heat shock transcription factor 1) and HSP70 versus control. They concluded that IPC induced miRNAs trigger cardioprotection similar to the delayed phase of IPC, possibly through upregulating eNOS, HSP70, and the HSP70 transcription factor HSF-1.

It should be noted that the study was largely observational without mechanistic insight into the target genes of the "IPC-miRNAs". The observed increases in the expression of eNOS, HSF-1 and HSP70 are unlikely ascribed to the upregulation of miR-1, miR-21 and miR-24; they are more like two parallel unrelated changes. Morever, it is not clear whether all the "IPC-miRNAs" or which one of the "IPC-miRNAs" really produce the protective effects. It may be more realistic to speculate that the "IPC-miRNAs" contain both "good" and "bad" miRNAs and the phenotyical protection is the balance between the "goods" and "bads". Further, uptake of miRNAs after myocardial injection was not verified.

miR-199a

Early hypoxia preconditioning (HPC) is an immediate cellular response to brief hypoxia/reoxygenation cycles that involves *de novo* protein, but not mRNA, synthesis. Hypoxia inducible factor (Hif)-1α is a well-established transcription factor that is rapidly induced by hypoxia through a post-transcriptional mechanism, in all tested cell types [Wang & Semenza, 1993]. It accounts for the transcription of 89% of genes that are upregulated during hypoxia [Greijer *et al.*, 2005]. In the heart, overexpression of Hif-1α during hypoxia resulted in a smaller infarct size following ischemia/reperfusion and was associated with higher capillary density, vascular endothelial growth factor, and inducible nitric oxide synthase (iNOS), in the periinfarct zone [Kido *et al.*, 2005].

Rane *et al* [2009] reported that miR-199a is miR-199a is sensitive to low oxygen levels and is acutely reduced to undetectable levels downregulated in cardiac myocytes of porcine hearts on a decline in oxygen tension. miR-199a directly targets and inhibits translation of Hif-1α mRNA during normoxia. This reduction is thus required for the rapid upregulation of its target, hypoxia-inducible factor (Hif)-1α. Replenishing miR-199a during hypoxia inhibits Hif-1α expression and its stabilization of p53 and, thus, reduces apoptosis. On the other hand, knockdown of miR-199a during normoxia results in the upregulation of Hif-1α and Sirtuin 1 (Sirt1) and reproduces hypoxia preconditioning. Sirt1 is also a direct target of miR-199a and is responsible for downregulating prolyl hydroxylase 2, required for stabilization of Hif-1α. Thus, we conclude that miR-199a is a master regulator of a hypoxia-triggered pathway and can be exploited for preconditioning cells against hypoxic damage. In addition, the data demonstrate a functional link between 2 key molecules that regulate hypoxia preconditioning and longevity.

REFERENCES

Ai J, Zhang R, Li Y, Pu J, Lu Y, Jiao J, Li K, Yu B, Li Z, Wang R, Wang L, Li Q, Wang N, Shan H, Li Z, Yang B. (2010) Circulating microRNA-1 as a potential novel biomarker for acute myocardial infarction. Biochem Biophys Res Commun 391:73–77.

Bolli R. (2000) The late phase of preconditioning. Circ Res 87:972–983.

Boštjančič E, Zidar N, Stajer D, Glavač D. (2009) MicroRNAs miR-1, miR-133a, miR-133b and miR-208 are dysregulated in human myocardial infarction. Cardiology 115:163–169.

Boštjančič E, Zidar N, Stajer D, Glavač D. (2010) MicroRNA miR-1 is up-regulated in remote myocardium in patients with myocardial infarction. Folia Biol (Praha) 56:27–31.

Greijer AE, van der Groep P, Kemming D, Shvarts A, Semenza GL, Meijer GA, van de Wiel MA, Belien JA, van Diest PJ, van der Wall E. (2005) Up-regulation of gene expression by hypoxia is mediated predominantly by hypoxia-inducible factor 1 (HIF-1). J Pathol 206:291–304.

Heusch G, Boengler K, Schulz R. (2008) Cardioprotection: nitric oxide, protein kinases, and mitochondria. Circulation 118:1915–1919.

Kido M, Du L, Sullivan CC, Li X, Deutsch R, Jamieson SW, Thistlethwaite PA. (2005) Hypoxia-inducible factor 1-alpha reduces infarction and attenuates progression of cardiac dysfunction after myocardial infarction in the mouse. J Am Coll Cardiol 46:2116 –2124.

Kloner RA. (1993) Does reperfusion injury exist in humans? J Am Coll Cardiol 21:537–545.

Mott JL, Kobayashi S, Bronk SF, Gores GJ. (2007) mir-29 regulates Mcl-1 protein expression and apoptosis. Oncogene 26:6133–6140.

Murry CE, Jennings RB, Reimer KA. (1986) Preconditioning with ischemia: a delay of lethal cell injury in ischemic myocardium. Circulation 74:1124–1136.

Rane S, He M, Sayed D, Vashistha H, Malhotra A, Sadoshima J, Vatner DE, Vatner SF, Abdellatif M. (2009) Downregulation of miR-199a derepresses hypoxia-inducible factor-1alpha and Sirtuin 1 and recapitulates hypoxia preconditioning in cardiac myocytes. Circ Res 104:879–886.

Tang Y, Zheng J, Sun Y, Wu Z, Liu Z, Huang G. (2009) MicroRNA-1 regulates cardiomyocyte apoptosis by targeting Bcl-2. Int Heart J 50:377–387.

Wang GK, Zhu JQ, Zhang JT, Li Q, Li Y, He J, Qin YW, Jing Q. (2010) Circulating microRNA: a novel potential biomarker for early diagnosis of acute myocardial infarction in humans. Eur Heart J 31:659–666.

Wang GL, Semenza GL. (1993) General involvement of hypoxia-inducible factor 1 in transcriptional response to hypoxia. Proc Natl Acad Sci USA 90:4304–4308.

Wynne AM, Mocanu MM, Yellon DM. (2005) Pioglitazone mimics preconditioning in the isolated perfused rat heart: a role for the prosurvival kinases PI3K and P42/44MAPK. J Cardiovasc Pharmacol 46:817–822.

Yang B, Lin H, Xiao J, Lu Y, Luo X, Li B, Zhang Y, Xu C, Bai Y, Wang H, Chen G, Wang Z. (2007) The muscle-specific microRNA miR-1 causes cardiac arrhythmias by targeting GJA1 and KCNJ2 genes. Nat Med 13:486–491.

Yasuda S, Kobayashi H, Iwasa M, Kawamura I, Sumi S, Narentuoya B, Yamaki T, Ushikoshi H, Nishigaki K, Nagashima K, Takemura G, Fujiwara T, Fujiwara H, Minatoguchi S. (2009) Antidiabetic drug pioglitazone protects the heart via activation of PPAR-gamma receptors, PI3-kinase, Akt, and eNOS pathway in a rabbit model of myocardial infarction. Am

J Physiol Heart Circ Physiol 2009;296:H1558–1565.

Ye Y, Hu Z, Lin Y, Zhang C, Perez-Polo JR. (2010) Down-regulation of microRNA-29 by antisense inhibitors and a PPAR-{gamma} agonist protects against myocardial ischemia-reperfusion injury. Cardiovasc Res. 2010 Feb 17. [Epub ahead of print]

Yin C, Wang X, Kukreja RC. (2008) Endogenous microRNAs induced by heat-shock reduce myocardial infarction following ischemia-reperfusion in mice. FEBS Lett 582:4137–4142.

Yin C, Salloum FN, Kukreja RC. (2009) A novel role of microRNA in late preconditioning: Upregulation of endothelial nitric oxide synthase and heat shock protein 70. Circ Res 104:572–575.

CHAPTER 6

miRNAs in Cardiac Arrhythmogenesis

Abstract: Cardiomyocytes are excitable cells that can generate and propagate excitations; excitability is a fundamental characteristic of these cells, which is reflected by action potential, the changes of transmembrane potential as a function of time, orchestrated by ion channels, transporters, and cellular proteins. The electrical excitation evoked in muscles must be transformed into mechanical contraction through the so-called excitation-contraction coupling mechanism, and the proper contraction of cardiac muscles then drives pumping of blood to the body circulation. Arrhythmias are electrical disturbances that can result in irregular heart beating with consequent insufficient pumping of blood. Arrhythmias are often lethal, constituting a major cause for cardiac death, particularly sudden cardiac death, in myocardial infarction and heart failure. Recent studies have led to the discovery of microRNAs (miRNAs) as a new player in the cardiac excitability by fine-tuning expression of ion channels, transporters, and cellular proteins, which determines the arrhythmogenicity in many conditions. This review article will give a comprehensive summary on the data available in the literature. The basics of cardiac excitability are first introduced, followed by a brief introduction to the basics of miRNAs. Then, studies on regulation of cardiac excitability by miRNAs are described and analyzed.

INTRODUCTION

Normal cardiac function relies on normal electrical activities of cardiomyocytes. Cardiomyocytes are excitable cells that can generate and propagate excitations; excitability is a fundamental characteristic of these cells. The electrical excitation evoked in muscles must be transformed into mechanical contraction through the so-called excitation-contraction coupling mechanism, and the proper contraction of cardiac muscles then drives pumping of blood to the body circulation. Arrhythmias are electrical disturbances that can result in irregular heart beating with consequent insufficient pumping of blood. Arrhythmias are often lethal, constituting a major cause for cardiac death, particularly sudden cardiac death, in myocardial infarction and heart failure and are one of the most difficult clinical problems.

Cardiac electrical activity is determined by a delicate balance among various ion channels, transporters, and their regulators. Intricate interplays of these ion channels maintain the normal heart rhythm. Dysfunction of any one of these proteins, due to functional impairment, expression deregulation, or genetic mutation, can break the balance rendering arrhythmias. Channelopathies, or dysfunction of the ion channels, which may result from genetic alterations in ion channel genes or aberrant expression of these genes, can render electrical disturbances predisposing to cardiac arrhythmias. Important to note is that altered miRNA expression can render expression deregulation of the proteins. This is because ion channel genes contain binding sites for miRNAs, which makes them potential targets for miRNAs and altered expression of miRNAs can either down- or up-regulate these genes.

BASIS OF ARRHYTHMOGENICITY

Cardiac Action Potential

The very basic electrical activity of the heart is represented by a cardiac action potential (AP). The cardiac AP reflects changes of cardiac transmembrane potential as a function of time. A typical cardiac AP contains a characteristic long plateau phase distinct from APs in non-cardiac cells. The properties of the cardiac AP are dynamically determined by underlying ionic currents generated by ion channels and ion transporters. A cardiac AP can be divided into 4 phases designated by the numbers 0 through 4, beginning with initial rapid depolarization (phase 0) and ending with the return to the resting state (phase 4).

In working atrial and ventricular muscle and specialized ventricular conducting tissue composed of Purkinje fiber cells, the cellular resting potential is set near the K^+ equilibrium potential of about -80 to -90 mV by the resting K^+ conductance, which is normally large because of a high resting permeability through inward rectifier K^+ current (I_{K1}). Upon activation, cells are depolarized by the rapid entry of Na^+ through Na^+ channels, generating a large inward-flowing (depolarizing) Na^+ current (I_{Na}). The maximum rate of voltage upstroke during phase 0 of the AP, dV/dt_{max}, is mainly determined by the size of I_{Na}. Ca^{2+} entry into the cell through the L-type Ca^{2+} current (I_{CaL}) also

Zhiguo Wang

contributes to phase 0 depolarization. This membrane depolarization is immediately followed by a brief rapid repolarization phase (phase 1) due to K^+ efflux through a rapidly activating and inactivating transient outward K^+ current (I_{to}). In atrial myocytes, another K^+ current named ultra-rapid delayed rectifier (I_{Kur}) is activated in accompany with I_{to}. Then cardiac cells enter the characteristic plateau phase (phase 2) during which there is a balance between inward currents I_{CaL} and I_{Na} and outward K^+ currents. During this phase there is progressive time-dependent sequential activations of delayed rectifier K^+ currents, first the rapid delayed rectifier K^+ current (I_{Kr}) and then the slow delayed rectifier K^+ current (I_{Ks}). I_{Kr} acts to terminate the AP with an appropriate delay by producing rapid phase 3 repolarization, and I_{Ks} serves as a repolarization reserve to prevent abnormal slowing of repolarization. Nodal type cells in the sino-atrial node and atrio-ventricular node maintain a more primitive phenotype, with a smaller resting K^+ conductance maintaining less negative resting potentials (further from the K^+ equilibrium potential) and a slower AP upstroke generated primarily by Ca^{2+} entry through the L-type Ca^{2+} channel. Typical examples of APs from various cardiac regions are illustrated in Fig. **1**.

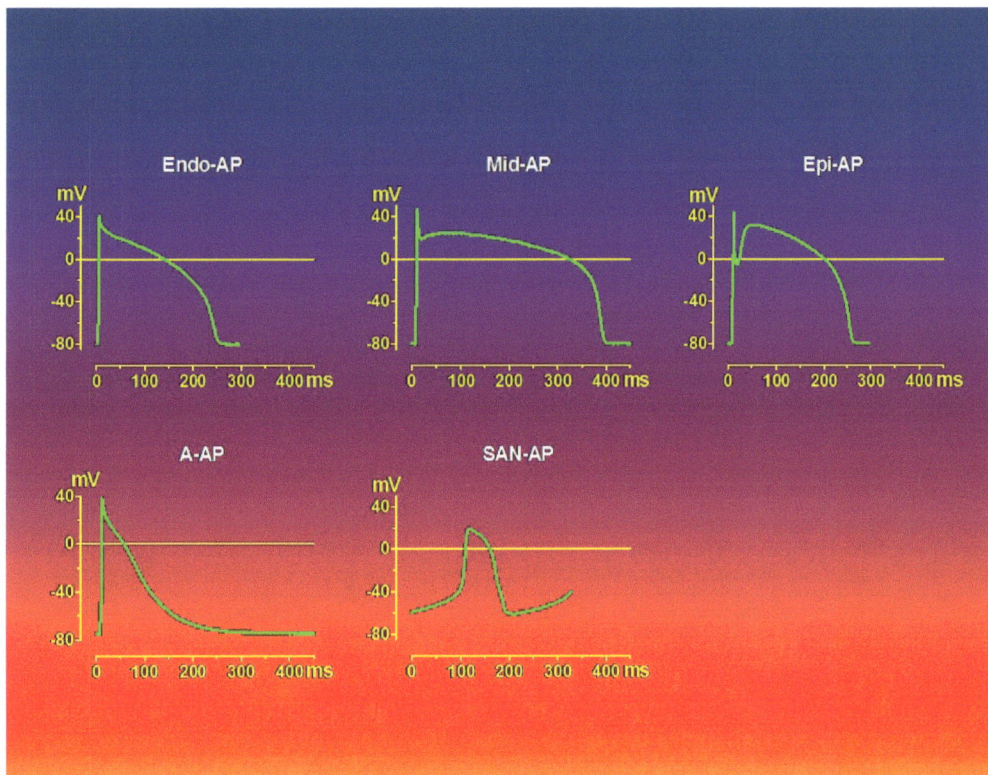

Figure 1: Cardiac action potential (AP) morphology recorded from various regions of a canine heart. Endo: left ventricular subendocardium; Mid: left ventricular midmyocardium; Epi: left ventricular subepicardium; A: right atrial free wall; SAN: sino-atrial nodal cell.

Cardiac Ion Channels

The electrical activities of the heart, as reflected by the action potential, are orchestrated by multiple categories of ion channels, the transmembrane proteins embedded across the cytoplasmic membrane of cardiomyocytes (Fig. **2**). The major ions flowing through ion channels in cardiac cells are sodium (Na^+), calcium (Ca^{2+}), and potassium (K^+). Different ion channels carry different ions; in other words, an ion channel is selectively permeable to a particular ion species. We therefore have Na^+ channels, Ca^{2+} channels, and K^+ channels in cardiac cells. The inward-flowing currents such as Na^+ and L-type Ca^{2+} currents tend to depolarize the membrane and delay repolarization to prolong action potential duration (APD), whereas the outward-going K^+ currents, mainly including I_{K1}, I_{to}, I_{Kr} and I_{Ks}, tend to repolarize membrane to shorten APD. Excessive slowing of repolarization due to inhibited outward currents or enhanced inward currents can cause abnormal APD prolongation leading to long QT type arrhythmias, whereas abnormal APD shortening favors re-entrant arrhythmias. Thus, abnormal membrane repolarization is an important aspect of cardiac excitability deregulation for arrhythmogenesis.

K^+ channels are represented by an extremely large and varied superfamily of genes including pore-forming α-subunits and regulatory auxiliary β-subunits. Based on the number of membrane-spanning domains, K^+ channel pore-forming α-subunits can be divided into three structural classes: channels with two, four or six transmembrane channels. Those with six transmembrane channels are further divided into six conserved families: voltage-gated (Kv), KQT, Eag, Slo, CNG and SK. In addition to their genetic and structural diversity, K^+ channels are also the most functionally diverse class of ion channel proteins. Many of these channels undergo remodeling processes, mostly with changed gene expressions, during pathogenesis. Thus, transcriptional regulatory regions of K^+ channel genes have been a subject attracting much research interest. A number of K^+ channels have been investigated on their genomic structures for transcriptional regulation, with their promoter regions identified and characterized. These include Kv1.4, Kv1.5, Kv3.1, Kv3.4, Kv4.2, Kv8.1, Kir1.1, Kir2.1, Kir3.1, Kir3.3, Kir6.2, Kir7.1, KCNQ2, etc. However, a majority of these studies have been conducted with rat and mouse genes; studies on promoters of human K^+ channel pore-forming α-subunits have been sparse despite that a recent report described the transcriptional control of several human KCNE genes (KCNE1-5) encoding a family of single-transmembrane-domain K^+ channel β-subunits that modulate the properties of several potassium channels [Lundquist *et al.*, 2006]. The present study is, to our knowledge, one of the three studies reported thus far on transcriptional control of human voltage-dependent K^+ channel pore-forming α-subunits; the other two are on Kv8.1 and KCNQ2 K^+ channel genes [Lundquist *et al.*, 2006; Xiao et a., 2001]. *Kv8.1* is located on chromosome 8q23.3, the susceptibility region for benign adult familial myoclonic epilepsy (BAFME) [Ebihara *et al.*, 2004], making this gene an interesting candidate for the disease. *KCNQ2* (20q13.3) is associated with benign familial neonatal convulsions (BFNC) [Xiao et a., 2001], a rare idiopathic epilepsy with an autosomal dominant mode of inheritance. *HERG1* (7q35-q36) is known to be responsible for type 2 long QT syndrome. However, how expressions of these genes affect the pathogenesis of the diseases remain elusive at this time.

Cardiac Excitability

The cardiac excitability is reflected by the action potential, involving at least five different aspects: cardiac conduction, membrane repolarization, automaticity, intracellular Ca^{2+} handling, and spatial heterogeneity of these above properties.

Cardiac Conduction

Conduction refers to the propagation of excitation within a cell and between cells, and cardiac conduction velocity is determined by the rate of membrane depolarization responsible for excitation generation and the intercellular conductance for excitation propagation. Na channel current (I_{Na}) is a critical determinant of membrane depolarization or the slope of the upstroke thereby the cardiac conduction.

However, for a whole heart to function properly, action potential must propagate smoothly throughout the heart from sinus node to atria then to ventricles through AV node. If any part of the conduction pathway is slowed or blocked, arrhythmias may occur. In this regard, connexion43 (Cx43) gap junction channel is of particular importance in ventricular muscles. Cx43 is critical for intercellular propagation of excitation. Impairment of these channels can cause abnormal cardiac conduction, the second aspect of cardiac excitability deregulation for arrhythmogenesis.

Cardiac Repolarization

After depolarization, the membrane potential begins to drop towards the resting state. The rate of membrane repolarization determines the length of action potential duration (APD) and effective refractory period (ERP) thereby the timeframe of availability for generation of a next excitation in a cardiac cell. The repolarization is governed by the delicate balance between the inward and the outward currents (Fig. **2**). The inward-flowing currents such as Na^+ and L-type Ca^{2+} currents tend to depolarize the membrane and delay repolarization to prolong action potential duration (APD), whereas the outward-going K^+ currents, mainly including I_{K1}, I_{to}, I_{Kr} and I_{Ks}, tend to repolarize membrane to shorten APD.

Cardiac Automaticity

Automaticity is a measure of the ease of cells to generate excitations or spontaneous membrane depolarization. Cardiac impulse or automaticity is normally generated in the sinus node, where the cells express abundant HCN genes, mainly HCN2 and HCN4, encoding *f*-channel proteins, as well as T-type Ca^{2+} channel. The current (I_f) generated by HCN channel proteins possesses the ability to depolarize cell membrane potential upon repolarization or hyperpolarization following the preceding AP. When the membrane potential reaches certain threshold, T-type Ca^{2+} channels activate and generate inward currents to elicit a spontaneous excitation. These genes are normally absent in the working muscles of adult heart; thus, muscles do not show automaticity and spontaneous activities.

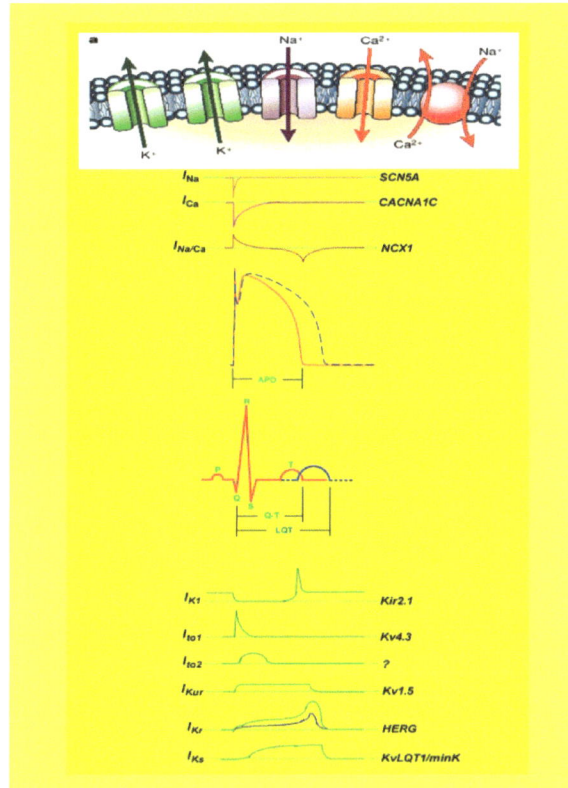

Figure 2: Diagram showing comparison between a ventricular action potential and electrocardiogram (ECG), along with the ion currents activated during various phases of AP and genes encoding the ion current channels.

Cardiac Intracellular Calcium Handling

In cardiac cells, several ion channels, transporters, and intracellular proteins coordinate to maintain proper intracellular Ca^{2+} handling. Normally, during an excitation Ca^{2+} enters cardiomyocytes via L-type Ca^{2+} current (I_{CaL}). This triggers the release of additional Ca^{2+} from sarcoplasmic reticulum (SR) Ca^{2+} stores through closely coupled SR Ca^{2+} release channel RyR2, commonly called Ca^{2+}-induced Ca^{2+} release. Ca^{2+} release channel function is regulated by phosphorylation, which depends particularly on key intracellular phosphorylating enzymes like protein kinase A (PKA) and Ca^{2+}/calmodulin-dependent protein kinase II (CaMKII), as well as a variety of phosphatases which cause dephosphorylation. Sarcoplasmic reticulum Ca^{2+} content depends on cellular Ca^{2+} entry, particularly via I_{CaL}, Ca^{2+} removal from the cell (particularly via the sarcolemmal Ca^{2+} pump and forward-mode Na^+-Ca^{2+} exchange) and Ca^{2+} pumping into the SR by the Ca^{2+}-ATPase Ca^{2+} pump (encoded by SERCA2a) [Nattel *et al.*, 2007].

Cardiac Spatial Heterogeneity of Excitability

Since the beginning of the last decade, when Day, McComb, and Campbell [1990] introduced the concept of QT dispersion as a potential marker of arrhythmogenicity risk and hence of cardiovascular morbi–mortality, various electrocardiographic ventricular repolarization parameters have been tested as prognostic factors in several conditions such as coronary heart disease, heart failure, cardiomyopathies, as well as in population-based studies. QT dispersion mainly reflects regional dispersion of ventricular repolarization or APD. Transmural differences across the ventricular wall have been well established in mammalian hearts of various species including human, dog, pig, feline, guinea pig and rabbit [Verduyn *et al.*, 1997a & 1997b; Volders *et al.*, 1999a & 1999b; Liu & Antzelevitch, 1995; Szentadrassy *et al.*, 2005; Gintant, 1995]. Action potentials from subepicardium (Epi) and midmyocardium (Mid) are characterized by spike-and-dome morphology which is absent in subendocardium (Endo), and APD is in general in the order of Mid>>Endo≥Epi. These transmural differences exist in both left (LV) and right (RV) ventricles with a similar gradient pattern [].[60] Not until recently, important interventricular heterogeneity of repolarization has also been recognized; APD is significantly shorter in RV than in LV [].[60-63] Under normal physiological conditions, these regional differences are genetically programmed with a certain pattern with APD gradient from long to short following the sequence of activation of myocardial mass during an excitation, which constitutes an intrinsic protection mechanism against arrhythmias which could be induced due to radial and retrograde excitation conduction. Under pathological situations, the spatial heterogeneity is abnormally increased and the intrinsic pattern of spatial heterogeneity may also be broken. These changes render the heart a loss of the intrinsic antiarrhythmic mechanism and a vulnerability to arrhythmogenesis. For example, enlarged interventricular dispersion has been shown to cause acquired *torsade de pointes* [].[61-62]

The spatial heterogeneity of cardiac repolarization is largely due to diversity and varying densities of repolarizing K^+ currents. The action potentials of Epi and Mid cells display a prominent Ca^{2+}-independent transient outward K^+ current (I_{to}) which is minimal in Endo cells. Mid cells are distinguished from the other cell types in that they display a smaller I_{Ks}, whereas I_{Kr} and I_{K1} appear similar in the three transmural layers [].[54] In canine hearts, distinct distributions of I_{to} and I_{Ks} have been reported between the ventricles, in which both currents were larger in RV than in LV [].[63] This difference could explain, at least in part, why APD is longer and why the APD-rate relationship is steeper in LV than in RV Mid cells. In the same study, some 30% differences of I_{Kr} tail current densities between two chambers at a plateau range of membrane potentials, with I_{Kr} larger in RV than in LV, were observed though the differences did not reach statistical significance due to relatively large variations. This amount of differences is actually fairly similar to our finding in rabbits.

Mechanisms for Arrhythmogenesis

Arrhythmias can occur when there is abnormality in the electrical activities or cardiac excitability: cardiac conduction, repolarization, automaticity, intracellular handling, or spatial heterogeneity of excitability.

Slowing of conduction, consequent to malfunction and/or downregualtion of Na^+ channel and/or Cx43 proteins, can result in unidirectional block of cardiac excitation conduction and shortening of wavelength (the product of effective refractory period and conduction velocity which defines the shortest circuit size that can sustain reentry) leading to reentrant arrhythmias [Nattel *et al.*, 2007].

Excessive slowing of repolarization due to inhibited outward currents or enhanced inward currents can cause abnormal APD prolongation leading to long QT type arrhythmias due to occurrence of early afterdepolarizations (EADs). On the other hand, abnormal APD shortening favors re-entrant arrhythmias and short Q-T syndrome. Thus, abnormal membrane repolarization is an important aspect of cardiac excitability deregulation for arrhythmogenesis.

Calcium channels account for excitation-contraction coupling and contribute to the plateau duration of AP. Intracellular Ca^{2+} homeostasis is also crucial to cardiac rhythmic activities. Ca^{2+} overload is a causal actor for arrhythmogenesis.

Pacemaker channels, which carry the non-selective cation current, are critical in generating sinus rhythm under normal conditions. Under certain circumstances, however, other regions of the heart can also produce spontaneous activities leading to focal excitations that can predispose to more severe arrhythmias. This is partially due to enhanced I_f the funny current and/or T-type Ca^{2+} current to enhance spontaneous membrane depolarization, as a result of re-expression of the genes that encode the *f*-channel proteins or T-type Ca^{2+} channel. The focal excitation or ectopic beat is often caused by these channels.

REGULATION OF CARDIAC EXCITABILITY BY miRNAs

To shed light on the role of miRNAs in regulating cardiac excitability, we conducted bioinformatics analysis using miRNA database websites miRBase and miRANDA to predict the potential target genes encoding cardiac ion channel proteins for the muscle-specific miRNAs *miR-1*, *miR-133* and *miR-208*. Based on the prediction, we then carried out luciferase reporter activity assays to verify the candidate targets. From these studies, we are able to obtain an overall picture of expression regulation of ion channel genes by the muscle-specific miRNAs and to come up with the following points. First, while *miR-1* and *miR-133* both can regulate multiple genes that cover nearly all cardiac ion channel subunits, *miR-208* does not seem to have any ion channel genes as its target for post-transcriptional repression. Second, *miR-1* or *miR-133* each has its distinct set of ion channel targets without overlapping genes for the other, except for the pacemaker channel gene *HCN2* that is regulated by both *miR-1* and *miR-133*. Third, some computationally predicted targets failed to be experimentally verified but some non-predicted genes were found to be targets for *miR-1* or *miR-133* according to our experimental results.

Intriguingly, in addition to the muscle-specific miRNAs, other miRNAs that are altered in their expression in ischemic myocardium and hypertrophic hearts may also be able to regulate the expression of cardiac ion channel genes. According to our bioinformatics prediction, *let-7f, miR-23a, miR-29a, miR-30, miR-124a, miR-125b, miR-150, miR-193, miR-214, miR-185, miR-494, miR-320* and *miR-351* all have cardiac ion channel genes as their putative targets. For instance, *miR-30* that is remarkably increased in its transcript level in myocardial infarction and decreased in cardiac hypertrophy [van Rooij *et al.*, 2006] can theoretically regulate several cardiac ion channel genes including *GJA1* (encoding connexin 43), *CACNB2* (dihydropyridine-sensitive L-type, calcium channel β2 subunit), and *KCNJ3* (Kir3.1 or GIRK1, a subunit of ACh-sensitive K^+ channel). *miR-195*, which was found to be among the most up-regulated miRNAs in cardiac hypertrophy [van Rooij *et al.*, 2006], is predicted to target *SCN5A* (encoding cardiac Na^+ channel α-subunit), *KCNJ2* (encoding Kir2.1, a pore-forming α-subunit of inward rectifier K^+ channel), and *KCNAB1* (β1-subunit of *Shaker*-type voltage-gated K^+ channels).

These theoretical analyses prompted us to carry out experimental investigations to link miRNAs to ion channel genes.

Regulation of Cardiac Conduction

The first evidence for the role of miRNAs in controlling cardiac excitability through targeting ion channel genes came from our study on acute myocardial ischemia (AMI) in 2007 [Yang *et al.*, 2007]. We found that *miR-1* is overexpressed (~2.8 fold increase) in the myocardium of individuals with coronary artery disease (CAD) relative to healthy hearts. To explore the mechanisms, we used a rat model of myocardial infarction induced by occlusion of the left anterior descending coronary artery for 12 hours that corresponds to the peri-infarction period during which phase II ischemic arrhythmias often occur, which represents a major challenge to our understanding and management of the disorder [Clements-Jewery *et al.*, 2005]. We found a similar increase (~2.6-fold) in miR-1 expression in the ischemic hearts of rat, which is accompanied by exacerbated arrhythmogenesis [Yang *et al.*, 2007]. Elimination of miR-1 by an antisense inhibitor in infarcted rat hearts relieved arrhythmogenesis. miR-1 overexpression slows cardiac conduction, as indicated by the widening of QSR complex and actual measurement of cardiac velocity, and depolarizes the cytoplasmic membrane, which is likely the cellular mechanism for the arrhythmogenic potential of miR-1. We further established GJA1, which encodes Cx43, and KCNJ2, which encodes the K^+ channel subunit Kir2.1 [Wang *et al.*, 1998], as target genes for miR-1. Cx43 is critical for inter-cell conductance and Kir2.1 for setting and maintaining membrane potential. Repression of these proteins by miR-1 explains for miR-1-induced slowing of cardiac conduction. We therefore proposed that myocardial infarction upregulates miR-1 expression via some unknown factors, which induces post-transcriptional repression of GJA1 and KCNJ2, resulting in conduction slowing leading to ischemic arrhythmias.

A separate study from Srivastava's group in the same time [Zhao *et al.*, 2007] demonstrated that targeted deletion of miR-1-2 in mice resulted in 50% lethality that was largely attributable to ventricular–septal defects. However, approximately half of the surviving mutant animals experienced electrophysiological defects, mainly conduction block, and sudden death. These authors determined *in vivo miR-1-2* targets, including the cardiac transcription factor, Irx5, which represses KCND2, a potassium channel subunit (Kv4.2) responsible for transient outward K^+ current (I_{to}) and is thereby critical for maintaining the ventricular repolarization gradient [Costantini *et al.*, 2005]. Their study suggests that the combined loss of Irx5 and Irx4 disrupted ventricular repolarization with a predisposition to arrhythmias. The increase in Irx5 and Irx4 protein levels in *miR-1-2* mutants corresponded well with a decrease in KCND2 expression. Clearly, loss-of-function of *miR-1* and Dicer mutant embryos affect conductivity through K^+ channels which supports a central role for *miR-1* for fine tuning the regulation of cardiac electrophysiology in pathological and normal conditions.

Apparently, a proper concentration range of miR-1 in myocardium is required for maintaining normal cardiac conduction; either excessive decrease or increase in miR-1 level can induce arrhythmia.

Regulation of Cardiac Repolarization

Cardiac repolarization is determined by an intricate interplay and a delicate balance between inward and outward ion currents. The rate of repolarization importantly affects the likelihood of arrhythmogenesis and regional dispersion of ventricular repolarization is a marker of arrhythmogenicity risk. The spatial heterogeneity of cardiac repolarization is largely due to diversity and varying densities of repolarizing K^+ currents. Slowly activating delayed rectifier K^+ current (I_{Ks}) along with its underlying channel proteins KCNQ1 (pore-forming α-subunit) and KCNE1 (auxiliary β-subunit) importantly affect cardiac APD and arrhythmogenesis through two mechanisms. First, I_{Ks} acts as a powerful repolarization reserve or safety factor to restrict excessive cardiac APD and QT prolongation caused by other factors. Removal of this safety factor facilitates LQTS [Roden & Yang, 2005]. Second, distribution of I_{Ks} in the heart follows important spatial patterns in at least four different axes: (1) transmural heterogeneity with epicardium (Epi)≥endocardium (Endo)>midmyocardium (Mid) [Liu & Antzelevitch, 1995; Gintant, 1995; Szabo *et al.*, 2005], (2) interventricular gradient with right ventricle (RV)>left ventricle (LV) [Volders *et al.*, 1999a; Ramakers *et al.*, 2003], transseptal gradient with RV septum>LV septum [Ramakers *et al.*, 2005], and apex-base difference with apical area>basal area [Szentadrassy *et al.*, 2005]. These intrinsic spatial patterns of distribution are important in maintaining the sequential excitations of cardiac muscles, and disruption of the patterns and/or exaggeration of the regional heterogeneity can create substrates for arrhythmogenesis. We have experimentally established *KCNQ1* and *KCNE1*, the long QT syndrome genes, as targets for repression by *miR-133* and *miR-1*, respectively [Luo *et al.*, 2007]. More importantly, we found that the distribution of *miR-133* and *miR-1* transcripts within the heart is also spatially heterogeneous with the patterns corresponding to the spatial distribution of KCNQ1 and KCNE1 proteins and I_{Ks}. Specifically, the *miR-133* level was found Base>Apex and Mid>Epi, a pattern reciprocal to the regional distribution of KCNQ1 proteins. The same logic can be applied to *miR-1* and KCNE1; the characteristic regional distributions of *miR-1*, Base>Apex and Epi>Mid, can be one of the causal factors for the converse transmural and apical-basal gradients of KCNE1 protein levels. *miR-1* and *miR-133* do not show any chamber-dependent differences. Our study thus revealed that the muscle-specific miRNAs miR-1 and miR-133 not only repress the K^+ channel genes but also regulate the spatial heterogeneity of expression of these ion channels thus the arrhythmogenesis associated with abnormal repolarization and regional dispersion of excitability.

Abnormal QT interval prolongation is a prominent electrical disorder and has been proposed a predictor of mortality in patients with diabetes mellitus, presumably because it is associated with an increased risk of sudden cardiac death consequent to lethal ventricular arrhythmias. In a subsequent study, we identified *ether-a-go-go* related gene (ERG), another long QT syndrome gene encoding a key K^+ channel (I_{Kr}) in cardiac cells, as a target for miR-133 [Xiao *et al.*, 2007]. We have previously found that the *ether-a-go-go* related gene (ERG), is severely depressed in its expression at the protein level but not at the mRNA level in diabetic subjects [Zhang *et al.*, 2006]. The reduced protein level of ERG is a causal factor for the abnormal QT prolongation in diabetic hearts. In an effort to understand the mechanisms underlying the disparate alterations of ERG protein and mRNA, we performed a study on expression regulation of ERG by miRNAs in a rabbit model of diabetes [Xiao *et al.*, 2007]. We found a remarkable overexpression of *miR-133* in diabetic hearts and in parallel, the expression of serum response factor (SRF), a transactivator of *miR-133*, was also found robustly increased. Delivery of exogenous *miR-133* into the rabbit myocytes and cell lines produced posttranscriptional repression of ERG, down-regulating ERG protein level without altering its transcript level. Correspondingly, forced expression of *miR-133* also caused substantial depression of I_{Kr}, an effect abrogated by the *miR-133* antisense inhibitor. Functional inhibition or gene silencing of SRF down-regulated *miR-133* expression and increased I_{Kr} density. Repression of ERG by *miR-133* likely underlies the differential changes of ERG protein and transcript thereby depression of I_{Kr} in diabetic cardiomyopathy.

Together these results, it appears that miR-133 targets the two most important long QT syndrome genes KCHN2 and KCNQ1, the mutations of which account for >80% genetic LQTS and a majority of acquired LQTS as well. This property of miR-133 bears significant physiological implications as revealed by the subsequent studies described below.

Inhibition of individual K^+ currents can cause functionally based compensatory increases in other K^+ currents that minimize excessive changes in action potential duration, a phenomenon known as repolarization reserve. To test the possibility that sustained K^+ channel inhibition may induce remodeling of ion current expression, Nattel's laboratory studied adult canine left ventricular cardiomyocytes incubated in primary culture and paced at a physiological rate (1 Hz) for 24 hours in the presence or absence of the highly selective rapid delayed-rectifier K^+ current (I_{Kr}) blocker dofetilide (5 nM). Sustained dofetilide exposure led to shortened action potential duration and increased repolarization reserve (manifested as a reduced action potential duration–prolonging response to I_{Kr} blockade). These repolarization changes were accompanied by increased slow delayed-rectifier (I_{Ks}) density, whereas I_{Kr}, transient-outward (I_{to}), inward-rectifier (I_{K1}), L-type Ca^{2+} (I_{CaL}), and late Na^+ current remained unchanged. The mRNA expression corresponding to KvLQT1 and minK was unchanged, but their protein expression was increased, suggesting post-transcriptional regulation. Strikingly, the expression levels of *miR-133a* and *miR-133b* were significantly decreased in cells cultured in dofetilide compared with control, possibly accounting for KvLQT1 protein upregulation. These results suggest that feedback control of ion channel expression by miRNAs may influence repolarization reserve. Similar feedback control of ion channel expression has also been observed in other occasions. For example, in transgenic mice engineered to lack the rapid transient-outward K^+ current (I_{tof}), evidence exists for compensatory upregulation of the slow transient outward component (I_{tos}) carried by Kv1.4 subunits [Guo *et al.*, 1999 & 2000]. Similarly, in mice lacking $I_{K,slow1}$ carried by Kv1 subunits, an apparent compensatory regulatory response of Kv2.1 subunits to increase $I_{K,slow2}$ can be observed [Zhou *et al.*, 2003].

In a most recent study, Matkovich *et al* [2010] described that miR-133a is downregulated in transverse aortic constriction (TAC) and isoproterenol-induced hypertrophy in mice. Using *MYH6* (βMHC) promoter-directed expression of a miR-133a genomic precursor, increased cardiomyocyte miR-133a (by 13 folds) had no effect on postnatal cardiac development assessed by measures of structure, function, and mRNA profile. However, increased miR-133a levels increased QT intervals in surface electrocardiographic recordings and action potential durations in isolated ventricular myocytes, with a decrease in the fast component of the transient outward K^+ current, $I_{to,f}$, at baseline. Transgenic expression of miR-133a prevented TAC-associated miR-133a downregulation and improved myocardial fibrosis and diastolic function without affecting the extent of hypertrophy. $I_{to,f}$ decrease normally observed post-TAC was prevented in miR-133a transgenic mice, although action potential duration and QT intervals did not reflect this benefit. miR-133a transgenic hearts had no significant alterations of basal or post-TAC mRNA expression profiles, although decreased mRNA and protein levels were observed for the $I_{to,f}$ auxiliary KChIP2 subunit, which is not a predicted target. These findings suggest that regulation of *Kcnip2* transcript levels in miR-133a transgenic hearts occurs by an indirect mechanism. Though we have characterized the regulation of I_{Kr} and I_{Ks} encoding genes by miR-133 in rabbits and canines [Xiao *et al.*, 2007; Xiao *et al.*, 2009; Luo *et al.*, 2007], this study did not observe these changes because the contribution of these currents to action potential repolarization in the intact mouse ventricle is negligible [Babij *et al.*, 1998].

Regulation of Cardiac Automaticity

The pacemaker current I_f, carried by the hyperpolarization-activated channels encoded mainly by the *HCN2* and *HCN4* genes in the heart, plays an important role in rhythmogenesis. Their expressions reportedly increase in hypertrophic and failing hearts, contributing to arrhythmogenicity under these conditions [Fernandez-Velasco *et al.*, 2003; Stilli *et al.*, 2001]. We performed a study on post-transcriptional regulation of HCN2 and HCN4 by miRNAs, experimentally establishing *HCN2* mRNA as a target for repression by the muscle-specific miRNAs *miR-1* and *miR-133*, and *HCN4* as a target for *miR-1* only [Luo *et al.*, 2007]. We further unraveled robust increases in *HCN2/HCN4* transcripts and protein levels in a rat model of left ventricular hypertrophy and in angiotensin II-induced neonatal cardiomyocyte hypertrophy. The upregulation of HCN2/HCN4 was accompanied by reduction of *miR-1/miR-133*. Overexpression of *miR-1/miR-133* by transfection prevented largely overexpression of *HCN2/HCN4* in hypertrophic cardiomyocytes. Our data indicate that *miR-1/miR-133* act to limit overexpression of HCN2/HCN4 at protein level and downregulation of *miR-1/miR-133* underlies partially the abnormal enhancement of HCN2/HCN4 expression in hypertrophic hearts.

Regulation of Intracellular Calcium Handling

The arrhythmogenic potential of miR-1 was reproduced by a recent study in rat [Terentye *et al.*, 2009], but a mechanism related to intracellular Ca^{2+} handling was identified. Alterations in intracellular Ca^{2+} cycling have been implicated in different cardiac diseases, including arrhythmia and heart failure. Cardiac contractility is modulated by reversible phosphorylation of the components of SR Ca^{2+} release machinery, including the L-type Ca^{2+} channel (dihydropyridine receptor [DHPR]), RyR2, and PLB, protein kinase (PKA) [Bers, 2002; Wehrens & Marks, 2004], Ca^{2+}/calmodulin-dependent protein kinase (CaMKII) [Maier & Bers, 2007; Grueter *et al.*, 2007], and phosphatases PP1 and PP2A [Steenaart *et al.*, 1992; Davare *et al.*, 2000; Marx *et al.*, 2000]. The PP2A catalytic subunit has been shown to complex with DHPR and RyR2 and is critical to dephosphorylation of these proteins following their phosphorylation by PKA and/or CaMKII [Davare *et al.*, 2000; Marx *et al.*, 2000]. Consistent with the role of B56α in conveying PP2A phosphatase activity to DHPR, B56α has been shown to coimmunoprecipitate with the α-subunit of the cardiac L-type Ca^{2+} channel (Cav1.2) [Hall *et al.*, 2006].

Terentye *et al* [2009] investigated the effects of increased expression of miR-1 on excitation–contraction coupling and Ca^{2+} cycling in rat ventricular myocytes using methods of electrophysiology, Ca^{2+} imaging and quantitative immunoblotting. Adenoviral-mediated overexpression of miR-1 in myocytes resulted in a marked increase in the amplitude of the inward Ca^{2+} current, flattening of Ca^{2+} transients voltage dependence, and enhanced frequency of spontaneous Ca^{2+} sparks while reducing the sarcoplasmic reticulum Ca^{2+} content as compared with control. In the presence of isoproterenol, rhythmically paced, miR-1–overexpressing myocytes exhibited spontaneous arrhythmogenic oscillations of intracellular Ca^{2+}, events that occurred rarely in control myocytes under the same conditions. The effects of miR-1 were completely reversed by the CaMKII inhibitor KN93. Although phosphorylation of phospholamban was not altered, miR-1 overexpression increased phosphorylation of RyR2 at S2814 for CaMKII but not at S2808 for PKA. Overexpression of miR-1 was accompanied by a selective decrease in expression of the protein phosphatase PP2A regulatory subunit B56α involved in PP2A targeting to specialized subcellular domains. It is suggested that miR-1 enhances cardiac excitation–contraction coupling by selectively increasing phosphorylation of the L-type and RyR2 channels via disrupting localization of PP2A activity to these channels. miR-1 enhanced the functional activity of RyR2 channels and thus resulted in increased EC coupling gain, elevated diastolic SR Ca^{2+} leak, and reduced SR Ca^{2+} content and promoted arrhythmogenic disturbances in myocyte Ca^{2+} cycling [Terentye *et al.*, 2009].

Regulation of Spatial Heterogeneity of Excitability

Day, McComb, and Campbell [Day *et al.*, 1990] first introduced the concept of QT dispersion as a potential marker of arrhythmogenicity risk, reflecting regional dispersion of ventricular repolarization. For example, enlarged interventricular differences have been shown to cause acquired LQTS [Verduyn *et al.*, 1997a and 1997b]. The spatial heterogeneity of cardiac repolarization is largely due to diversity and varying densities of repolarizing K^+ currents. The well-recognized regional heterogeneity of I_{Ks} is one of the important factors determining the spatial dispersion of electrical activities [Liu & Antzelevitch, 1995].

The voltage-gated KCNQ1 (KvLQT1, Kv7.1) K^+ channels are expressed in a variety of tissues throughout the body and regulate key physiological functions. In cardiac myocytes, the KCNQ1 subunit assembles with the KCNE1 β-subunit (minK) to form a channel complex constituting the slow delayed rectifier current I_{Ks} (Barhanin *et al.*, 1996; Sanguinetti *et al.*, 1996). Mutations in KCNQ1 or KCNE1 can result in dysfunction of I_{Ks} and abnormality of cardiac repolarization, which is responsible for a majority of inherited long QT syndrome (LQTS). While the exact role of I_{Ks} in cardiac repolarization is still incompletely understood, it is clear that it can importantly affect cardiac action potential duration (APD) and arrhythmogenesis through two mechanisms. First, I_{Ks} acts as a powerful repolarization reserve or safety factor to restrict excessive cardiac APD and QT prolongation caused by other factors. Removal of this safety factor facilitates LQTS (Roden and Yang, 2005; Jost *et al.*, 2005). Second, distribution of I_{Ks} in the heart follows important spatial patterns in at least four different axes: (1) transmural heterogeneity with epicardium (Epi) ≥ endocardium (Endo) > midmyocardium (Mid) (Liu and Antzelevitch, 1995; Gintant, 1995; Szabo *et al.*, 2005), (2) interventricular gradient with right ventricle (RV) > left ventricle (LV) (Volders *et al.*, 1999a; Ramakers *et al.*, 2003), transseptal gradient with RV septum > LV septum (Ramakers *et al.*, 2005), and apex-base difference with apical area > basal area (Szentadrassy *et al.*, 2005). These intrinsic spatial patterns of distribution are important in maintaining the sequential excitations (depolarization and repolarization) of cardiac muscles, and disruption of the patterns and/or exaggeration of the regional heterogeneity can create substrates for arrhythmogenesis. Many pathological conditions can change the function and relative distribution of I_{Ks} by altering expression of its encoding genes (Ramakers *et al.*, 2003; Tsuji *et al.*, 2006; Volders *et al.*, 1999b;

Tsuji *et al.*, 2002; Akar and Tomaselli, 2005) Decreases in I_{Ks} density have been consistently observed in failing hearts of both animal models and patients. KCNQ1 and KCNE1 mRNA and protein levels were downregulated in both bradypaced and tachypaced rabbits. In chronic atrial-ventricular block, I_{Ks} is reduced likely due to transcriptional downregulation of KCNQ1 and KCNE1. Evidently, expression regulation of KCNQ1 and KCNE1 plays a critical role in defining the physiological function of I_{Ks}.

Based on computational predictions, neither KCNQ1 nor KCNE1 were considered targets for *miR-1/miR-133*, despite that they both have substantial complementarity to the miRNAs, likely due to unfavorable free energy status. However, by detailed analysis of the 3'UTR of *KCNQ1*, we identified four putative binding sites for *miR-133*. Similarly, the 3'UTR of *KCNE1* contains three putative *miR-1* binding sites. These multiple sites may cooperate with one another to confer significant effects of the miRNAs. Our experimental data convincingly demonstrated that the expression levels of KCNQ1 and KCNE1 are importantly regulated by *miR-133* and *miR-1*, respectively. Consistent with the principle of action of miRNAs, *miR-133* and *miR-1* decreased KCNQ1 and KCNE1 protein level without significantly affecting their mRNA levels. As to be discussed below, our study also revealed for the first time that distribution of the miRNAs in adult human hearts is heterogeneous and this heterogeneity may contribute to regional asymmetry of many proteins.

We have qualitatively reproduced the results reported in previous published studies: (1) KCNQ1 and KCNE1 distribute with significant interventricular gradients (RV>LV) at both mRNA and protein levels [Ramakers *et al.*, 2003], (2) The protein levels of both KCNQ1 and KCNE1 are higher in apical than in basal area [Szentadrassy *et al.*, 2005], despite that their mRNA levels are not significantly different, and (3) KCNQ1 protein level is greater in Epi than in Mid [Szabo *et al.*, 2005], whereas that of KCNE1 is the opposite, and there is no transmural difference in mRNA levels of KCNQ1 and KCNE1 [Pereon *et al.*, 2000]. Clearly, in addition to the regional difference of KCNQ1 and KCNE1 expressions, there is also an unparallel expression between the proteins and mRNAs of these genes. Intriguingly, the distribution patterns of Sp1 and *miR-1/miR-133* seem to provide a reasonable explanation for the observations. First, the interventricular difference of Sp1, which drives KCNQ1 and KCNE1 expressions, corresponds to the interventricular differences of KCNQ1 and KCNE1, at both mRNA and protein levels. By comparison, *miR-1* and *miR-133* do not show any chamber-dependent differences. These suggest that higher abundance of Sp1 in RV likely contributes to higher abundance of KCNQ1 and KCNE1 in the same chamber. Second, there is no difference of Sp1 expression either across the LV wall or along the apical-basal axis, coincident with the uniformity of distributions of KCNQ1 and KCNE1 transcripts in these two axes. Most strikingly, the *miR-133* level was found much greater in Base than in Apex and in Mid than in Epi. Considering the fact that *miR-133* represses KCNQ1 proteins leaving its transcripts unaltered, it is not difficult to understand why KCNQ1 protein levels have the opposite patterns of transmural and apical-basal gradients to those of *miR-133* whereas KCNQ1 mRNA has no transmural and apical-basal gradients. The same logic can be applied to *miR-1* and KCNE1; the characteristic regional distributions of *miR-1*, Base>Apex and Epi>Mid, can be one of the causal factors for the converse transmural and apical-basal gradients of KCNE1 protein levels. It seems reasonable to come to a conclusion that combination and interplay of the characteristic spatial distributions of Sp1 and *miR-1/miR-133* account, at least partially, for the regional heterogeneity of I_{Ks}-encoding genes and the disparity between protein and mRNA expressions of these genes. It should be mentioned, however, that our data are only supportive, but not confirmative, to this notion. Further studies using *in vivo* genetic engineering of Sp1 and *miR-1/miR-133* are absolutely needed to fully establish the link.

miRNAs AND ARRHYTHMIAS: EXPERIMENTAL EVIDENCE

Cardiac arrhythmias are a leading cause of mortality and morbidity in developed countries. Arrhythmias are electrical disturbances that can result in irregular heart beating. The hypothesis of altered excitability suggests that for arrhythmias to arise, the normal matrix must be perturbed by arrhythmogenic substrates to produce a proarrhythmic matrical condition to permit rhythmic disturbances caused by impaired excitation conduction/propagation, enhanced automaticity, or abnormal repolarization. Cardiac ion channels are fundamental determinants of cardiac excitability. Abnormalities of these ion channels, channelopathies, can be attributed to mutations in the genes encoding the channel proteins, which can predispose to arrhythmias. In many cases, malfunction of ion channels can also be ascribed to abnormally altered expression. Our findings that miRNAs regulate expression of cardiac ion channels strongly indicate a possibility of these miRNAs to influence arrhythmogenicity. The published studies have indeed generated data in support of this notion.

miRNAs and Ischemic Arrhythmias

One of the most deleterious alterations during myocardial infarction is the occurrence of ischemic arrhythmias. As already described, we have demonstrated that delivery of *miR-1* into the myocardium by *in vivo* gene transfer approach induces arrhythmias in otherwise healthy normal hearts and promotes arrhythmias including ventricular tachycardia and ventricular fibrillation in a rat model of myocardial infarction [Yang *et al.*, 2007]. We therefore proposed that myocardial infarction upregulates *miR-1* expression via some unknown factors, which induces post-transcriptional repression of *GJA1* and *KCNJ2*, resulting in conduction slowing leading to ischemic arrhythmias.

In addition, loss of cardiomyocytes following a myocardial infarction or in heart failure irreversibly damages the myocardium. One of the damages is an increased risk of arrhythmogenesis through creating discontinuity of excitation propagation and anisotropic conduction. One important form of cell death in these conditions is apoptosis, a programmed cell death. Our study [Xu *et al.*, 2007] and studies from other laboratories [Yu *et al.*, 2008; Tang *et al.*, 2009] showed that *miR-1* produced proapoptotic effects in cardiomyocytes in response to cellular stress such as oxidative stress, high glucose, and AMI. It is thus speculated that the proapoptotic action of *miR-1* may also a part of the mechanisms for its proarrhythmic effects.

Furthermore, intercellular conduction defects are also impacted by the increased interstitial fibrosis in the settings of MI, failing heart and atrial fibrillation, which can result in conduction discontinuity creating the substrate for arrhythmogenesis. For details on apoptosis and fibrosis, please see Chapters 12 and 13.

miRNAs and Diabetic QT Prolongation

Abnormal QT interval prolongation is a prominent electrical disorder and has been proposed a predictor of mortality in patients with diabetes mellitus, presumably because it is associated with an increased risk of sudden cardiac death consequent to lethal ventricular arrhythmias. ERG K^+ channels play a critical role in governing cardiac APD and impairment of ERG can cause substantial prolongation of APD favoring occurrence of early afterdepolarizaions (EADs), being the major cause of the acquired long QT syndrome. Repression of ERG by *miR-133* likely underlies depression of EAG function and contributes to repolarization slowing thereby QT prolongation and the associated arrhythmias in diabetic hearts [Xiao *et al.*, 2007]. In addition to ERG/I_{Kr}, we found that *miR-133* also repressed KCNQ1/I_{Ks} [Luo *et al.*, 2007; Xiao *et al.*. 2008]. It is interesting to see if miR-133 can create the substrate for LQTS by repressing these K^+ channels. In this regard, however, miR-133 has been found to be downregulated in most of the pathological conditions examined thus far [Carè *et al.*, 2007; Boštjančič *et al.*, 2009; Luo *et al.*, 2008]. This fact would imply that downregulation of miR-133 may be a protective mechanism against arrhythmogensis associated with repoilarization under these situations.

It should be noted that inhibition of ERG/I_{Kr} and KCNQ1/I_{Ks} by *miR-133* could also be antiarrhythmic under certain circumstances because the consequent prolongation of APD can lengthen ERP to reduce the likelihood of reentrant types of arrhythmias. This notion merits detailed studies.

miRNAs and Hypertrophic Automaticity

Repression of pacemaker channel gene *HCN2* and *HCN4* by *miR-1* and *miR-133* may confer their antiarrhythmic capability. *HCN2* is primarily expressed in ventricular myocytes where it can elicit ectopic heart beat leading to arrhythmias whereas *HCN4* is mainly distributed in the sinus nodal cells where it is critical for heart beat generation and heart rate regulation. Suppression of ectopic beat by suppressing HCN2 can antagonize arrhythmias and reduction of heart rate by suppressing HCN4 can minimize myocardial injuries during ischemia. In this regard, we can speculate that in myocardial infarction, increased *miR-1* level is expected to decrease HCN2 and HCN4 expression, and to limit enhancement of abnormal automaticity and occurrence of the associated arrhythmias. By contrast, in hypertrophic hearts, expression of *miR-1* and *miR-133* is reduced [Carè *et al.*, 2007; Luo *et al.*, 2008]. We have shown that this reduction contributes to the re-expression of HCN2/HCN4 and the enhanced automaticity [Luo *et al.*, 2008]. Under such conditions, arrhythmias arisen from ectopic beat may increase.

miRNAs and Atrial Fibrillation

Atrial fibrillation (AF) is a highly prevalent condition associated with pronounced morbidity and mortality; it can cause or exacerbate heart failure and is a risk factor for stroke. AF is characterized by the atrial electrical remodeling (or ion channel remodeling) process favoring the recurrence and maintenance of AF [Nattel, 2002; Nattel *et al.*, 2007]. A prominent finding in this remodeling is a shortening of effective refractory period (ERP) favoring reentry; this is primarily because of the shortening of the atrial action potential as a result of two critical changes. The first change is the reduction of L-type Ca^{2+} current (I_{CaL}) that serves to shorten the plateau duration. The second change is the increase in inward rectifier K^+ current (I_{K1}) [Bosch *et al.*, 1999; Cha *et al.*, 2004; Chen *et al.*, 2002; Dobrev *et al.*, 2001 & 2002; Gaborit *et al.*, 2005], a hallmark of atrial electrical remodeling in AF, which underlies the shortening of the terminal phase.

Girmatsion *et al* [2009] evaluated changes in miR-1 and Kir2 subunit expression in relation to I_{K1} alterations in left atrium (LA) of patients with persistent AF. Atrial tissue was obtained from 62 patients (31 with AF) undergoing mitral valve repair or bypass grafting. I_{K1} density was significantly increased in LA cells from patients with AF. There was a corresponding increase in Kir2.1 protein expression, but no change in other Kir proteins. Expression of inhibitory miR-1 was reduced by approximately 86% in tissue samples of AF patients. Kir2.1 mRNA was significantly increased. No change in Cx43 localization occurred. *Ex vivo* tachystimulation of human atrial slices up-regulated Kir2.1 and down-regulated miR-1, suggesting a primary role of atrial rate in miR-1 down-regulation and I_{K1} up-regulation. Because up-regulation of inward-rectifier currents is important for AF maintenance, these results provide potential new insights into molecular mechanisms of AF with potential therapeutic implications.

A large number of people smoke cigarettes and/or use over-the-counter nicotine products (patches and gums) to satisfy nicotine addiction. Serious, sometimes fatal, cases of atrial fibrillation (AF) have been reported in patients with and without detectable structural and/or history of heart disease from using a nicotine product, particularly when patients have smoked while using nicotine patches. In a study by Shan *et al* [2009] to elucidate molecular mechanisms underlying nicotine's promoting AF by inducing atrial structural remodeling, they found that nicotine stimulated remarkable collagen production and atrial fibrosis both *in vitro* in cultured canine atrial fibroblasts and *in vivo* in canine atrium subjected to rapid atrial pacing to induce AF. Nicotine produced significant upregulation of expression of TGF-β1 and TGFβRII at the protein level, and a 60–70% decrease in the levels of miRNAs miR-133 and miR-590. This downregulation of miR-133 and miR-590 partly accounts for the upregulation of TGF-β1 and TGFβRII, because our data established TGF-β1 and TGF-βRII as targets for miR-133 and miR-590 repression. Transfection of miR-133 or miR-590 into cultured atrial fibroblasts decreased TGF-β1 and TGF-βRII levels and collagen content. These effects were abolished by the antisense oligonucleotides against miR-133 or miR-590. The authors concluded that the profibrotic response to nicotine in canine atrium is critically dependent upon downregulation of antifibrotic miRNAs miR-133 and miR-590. This study also provides an evidence for the participation of non-ion channel proteins in arrhythmias.

miRNAs AND ARRHYTHMIAS: BIOINFORMATICS ANALYSIS

miRNAs are abundant non-coding mRNAs in terms of the species of miRNAs existing in a cell: to date, ~6400 vertebrates mature miRNAs have been registered in miRBase, an online repository for miRNA [Griffiths-Jones *et al.*, 2008], among which ~5100 miRNAs are found in mammals which include 718 human miRNAs. These miRNAs are predicted to regulates ~30% of protein–coding genes [Lewis *et al.*, 2005; Miranda *et al.*, 2006]. One common concern that somewhat subsides researchers' inner confidence on the published experimental data on miRNA-target interactions with high-level skeptics and thus hinders our understanding of the function of miRNAs is the possibility that a single protein-coding gene may be regulated by multiple miRNAs and *vice versa* an individual miRNA has the potential to target multiple protein-coding genes. For instance, in our previous study, *miR-1* was shown to target GJA1 (encoding gap junction channel protein connexin43) and KCNJ2 (encoding the Kir2.1 K^+ channel subunit) to cause slowing of cardiac conduction leading to ischemic arrhythmogenesis [Yang *et al.*, 2007]. However, it is conceivable that GJA1 and KCNJ2 are not the only ion channel targets for *miR-1*; it is also able to repress other genes such as SCN5A, CACNA1C, KCND2, KCNA5 and KCNE1 [4] and whether the repression of these genes other than GJA1 and KCNJ2 also contributes to the ischemic arrhythmogenesis remained unclear. On the other hand, GJA1 is predicted to be regulated by other miRNAs in addition to *miR-1* (including *miR-101, miR-125, miR-130, miR-19, miR-23*, and *miR-30*); whether these miRNAs are also involved in the deregulation of GJA1 in myocardial infarction remained unknown either. This same uncertainty or confusion expectedly exists in the interactions between literally all miRNAs and protein-coding genes. The only way to tackle this problem is the proper experimental approaches.

However, given the laborious nature of experimental validation of targets and the limited available experimentally validated data, computational prediction programs remain the only source for rapid identification of a putative miRNA target *in silico*. While currently available experimental approaches do not allow for thorough elucidation of the complete set of target genes of a given miRNA or of the complete array of mRNAs that regulate a given protein-coding genes, appropriate theoretical analyses might aid to resolve this intricate problem. To shed light on the issue, we performed a rationally designed bioinformatics analysis in conjunction with experimental approaches to identify the miRNAs from the currently available miRNA databases which have the potential to regulate human cardiac ion channel genes and to validate the analysis with several pathological settings associated with the deregulated miRNAs and ion channel genes in the heart [Luo *et al.*, 2010].

Initial Analysis of miRNAs with the Potential to Regulate Cardiac Ion Channel Genes

Our study was focused on the genes encoding cardiac cytoplasmic ion channels and electrogenic ion transporters. The list includes α- and β-subunits of Na^+ channels (SCN5A and SCN4B), α- and β-subunits of L-type Ca^{2+} channels (CACNA1C, CACNB1 and CACNB2), inward rectifier K^+ channel subunits (KCNJ2, KCNJ4, KCNJ12, KCNJ14 and KCNK1), voltage-gated K^+ channel pore-forming α-subunits (KCNA4, KCNA5, KCND2, KCND3, KCNH2, and KCNQ1), ACh-activated K^+ channel α-subunits (KCNJ3 and KCNJ5), ATP-sensitive K^+ channel α-subunit (KCNJ8) and receptor subunit (ABCC9), pacemaker hyperpolarization-activated cyclic-nucleotide gated cation channels (HCN2 and HCN4), gap junction channel proteins (GJA1, GJA5 and GJC1), transient receptor potential channel subunits (TRPC4 and TRPM4), chloride channel subunits (CLCN2, CLCN3 and CLCN6), K^+ channel β-subunits (KCNE1, KCNE2, KCHIP2 and KCNAB2), Na^+/Ca^{2+} exchanger NCX1 (SLC8A1), and Na^+/K^+-ATPase (ATP1A3 and ATP1B1). These cytoplasmic ion channels and electrogenic ion transporters play the fundamental roles in generating, maintaining and shaping cardiac electrical activities. Dysfunction of these proteins have been associated with a variety of pathological conditions of the heart.

As an initial "screening" process, we performed miRNA target prediction through the miRecords database [Xiao *et al.*, 2009]. This miRNA database integrates miRNA target predictions by 11 algorithms, as detailed in Methods section. Four of the 11 algorithms (microInspector, miTarget, NBmiRTar, and RNA22) were removed from our data analysis because they failed to predict; these websites require manual input of 3'UTR sequences of the genes. Thus, our data analysis was based upon the prediction from seven algorithms (TargetScan, DIANA-miT3.0, miRanada, PicTar, PITA, RNAHybrid, and miRTarget2) [38-44]. These prediction techniques are based on algorithms with different parameters (such as miRNA seed:mRNA 3'UTR complementarity, thermodynamic stability of base-pairing (assessed by free energy), evolutionary conservation across orthologous 3'UTRs in multiple species, structural accessibility of the binding sites, nucleotide composition beyond the seed sequence, number of binding sites in 3'UTR, and anti-correlation between miRNAs and their target mRNAs) and each of them are expected to provide a unique dataset. Some of them have higher sensitivity of prediction but low accuracy and the other weight on the accuracy in the face of reduced sensitivity. We collected all miRNAs predicted by at least four of the seven algorithms to have the potential to target any one of the selected cardiac ion channel and ion transporter genes. Meanwhile, we also collected all ion channel and ion transporter genes that contain the potential target site(s) (the binding site(s) with favorable free energy profiles) for at least one of the 718 human miRNAs. From the above two datasets, we noticed two points. First, out of 718 mature human miRNAs registered in miRBase, 429 miRNAs find their potential target site(s) in the 3'UTR(s) of at least one of the genes encoding cardiac ion channels and ion transporters. Second, all of the genes encoding cardiac ion channels and ion transporters selected for analysis, except for CLCN2, are the potential targets for miRNA regulation.

Detailed Analysis of the miRNAs with the Potential to Regulate Cardiac Ion Channel Genes

Using this cardiac miRNA expression profiling data in conjunction with published data obtained by real-time RT-PCR by Liang *et al* [2007], we refined the miRNA–target prediction by filtering out the miRNAs that are not expressed in the heart. In this way, we generated the modified datasets for subsequent analyses [Supplementary Table S2 & Table S3 in Luo *et al.*, 2010]. Detailed analysis of these two datasets revealed the following notes.

(1) One hundred ninety-three out of 718 registered human miRNAs or out of 222 miRNAs expressed in the heart have the potential to target the genes encoding human cardiac ion channels and transporters.

(2) Only two genes CLCN2 and KCNE2 were predicted not to contain the target site for miRNAs expressed in the heart.

(3) It appears that the most fundamental and critical ion channels governing cardiac excitability have the largest numbers of miRNAs for their regulators. These include SCN5A for I_{Na} (responsible for the upstroke of the cardiac action potential thereby the conduction of excitations), CACNA1C/CACNB2 for $I_{Ca,L}$ (accounting for the characteristic long plateau of the cardiac action potential and excitation-contraction coupling), KCNJ2 for I_{K1} (sets and maintains the cardiac membrane potential), SLC8A1 for NCX1 (an antiporter membrane protein which removes Ca^{2+} from cells), GJA1/GJC1 (gap junction channel responsible for intercellular conduction of excitation), and ATP1B1 for Na^+/K^+ pump (establishing and maintaining the normal electrochemical gradients of Na^+ and K^+ across the plasma membrane). Each of these genes is theoretically regulated by >30 miRNAs.

(4) The atrium-specific ion channels, including Kir3.4 for I_{KACh}, Kv1.5 for I_{Kur}, and CACNA1G for $I_{Ca,T}$, seem to be the rare targets for miRNAs (<5 miRNAs).

(5) All four genes for K^+ channel auxiliary □-subunits KCNE1, KCNE2, KCHiP, and KCNAB2 were also found to have less number of regulator miRNAs (<10).

(6) Intriguingly, 16 of these top 20 miRNAs are included in the list of the predicted miRNA-target dataset; the other four cardiac-abundant miRNAs *miR-21, miR-99, miR-100* and *miR-126* are in theory unable to regulate the genes for human cardiac ion channels and transporters.

(7) There is a rough correlation between the number of predicted targets and the abundance of miRNAs in the heart. It appears that the miRNAs within top 8 separate from the rest 12 less abundant miRNAs in their number of target genes (Fig. 1B). The muscle-specific miRNA *miR-1* was predicted to have the largest number of target genes (9 genes) among all miRNAs most abundantly expressed in the heart, followed by *miR-30a/b/c, miR-24* and *miR-125a/b* that have 6 target genes each. The muscle-specific miRNA *miR-133* has four target genes and three of them (KCNH2, KCNQ1 and HCN2) have been experimentally verified [Xiao *et al.*, 2007; Luo *et al.*, 2007 & 2008].

(8) Comparison of the target genes of the two muscle-specific miRNAs *miR-1* and *miR-133* revealed that they might play different role in regulating cardiac excitability. It appears that *miR-1* may be involved in all different aspects of cardiac excitability: cardiac conduction by targeting GJA1 and KCNJ2, cardiac automaticity by targeting HCN2 and HCN4, cardiac repolarization by targeting KCNA5, KCND2 and KCNE1, and Ca^{2+} handling by targeting SLC8A1. By comparison, *miR-133a/b* mainly controls cardiac repolarization through targeting KCNH2 (encoding HERG/I_{Kr}) and KCNQ1 (encoding KvLQT1/I_{Ks}), the two major repolarizing K^+ channels in the heart. The *let-7* seed family members seem to regulate mainly cardiac conduction by targeting SCN5A (Nav1.5 for intracellular conduction) and GJC1 (Cx45 for intercellular conduction). *miR-30a/b/c* and *miR-26a/b, miR-125a/b, miR-16,* and *miR-27a/b* were predicted to be L-type Ca^{2+} channel "blockers" through repressing α1c- and/or β1/β2-subunits (Fig. **3**).

Application of the Theoretical Analysis to Explaining the Electrical Remodeling Processes of Cardiac Diseases

Next, we tried to analyze whether the theoretical prediction could be applied to explaining some established observations of the electrical remodeling related to deregulation of both miRNAs and the genes for ion channels and transporters. Three pathological conditions, cardiac hypertrophy/heart failure, ischemic myocardial injuries, and atrial fibrillation, were studied because the participation of miRNAs in these conditions has previously been investigated.

Cardiac Hypertrophy and Heart Failure

The adult heart is susceptible to stress (such as hemodynamic alterations associated with myocardial infarction, hypertension, aortic stenosis, valvular dysfunction, etc) by undergoing remodeling process, including electrical/ionic remodeling. The remodeling process may originally be adaptive in nature, but is in the face of increased risk of arrhythmogenesis. The mechanisms for arrhythmogenesis in failing heart involve [46]: (1) Abnormalities in spontaneous pacemaking function (enhanced cardiac automaticity) as a result of increases in atrial and ventricular I_f due to increased expression of HCN4 channel may contribute to ectopic beat formation in CHF; (2) Slowing of cardiac repolarization thereby prolongation of APD due to reductions of repolarizing K^+ currents (including I_{K1}, I_{Ks}, and I_{to1}) provides the condition for occurrence of early afterdepolarizations (EADs) leading to triggered activities; (3) Delayed afterdepolarizations (DADs) due to enhanced Na^+-Ca^{2+} exchanger (NCX1) activity in cardiac hypertrophy/CHF is a consistent finding by numerous studies. Upregulation of NCX1 expression is the major cause for the enhancement; (4) Reentrant activity due to slowing of cardiac conduction velocity.

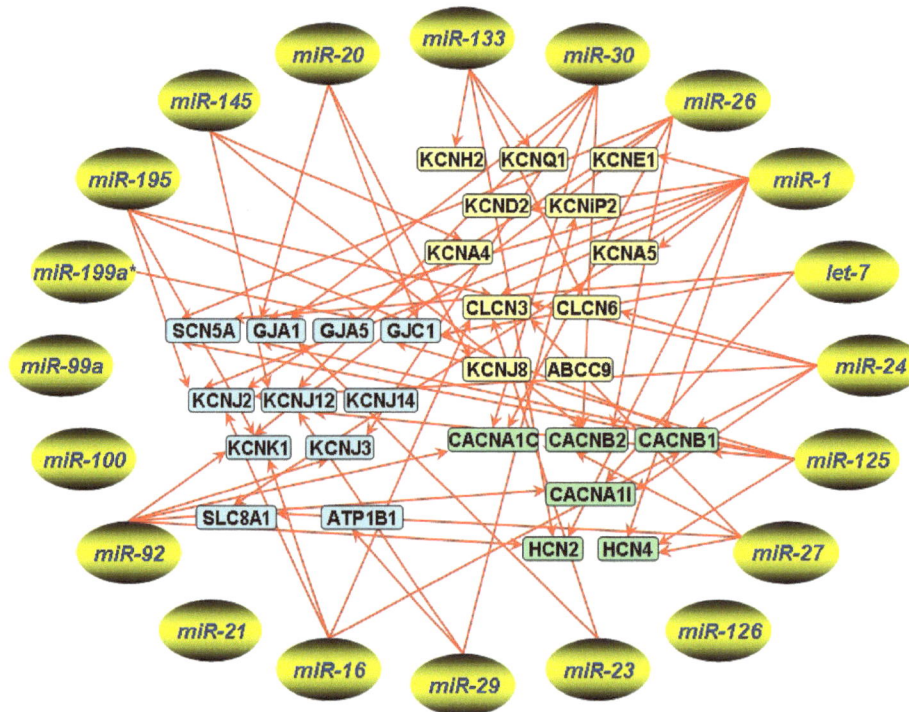

Figure 3: Predicted gene targeting of the top 20 most abundantly expressed miRNAs in myocardium. The arrows indicate repression of the genes by the connected miRNAs. The target genes are roughly divided into groups in three different colors: the genes for cardiac conduction in blue, the genes for cardiac automaticity in green, and the genes for cardiac repolarization in yellow.

To date, there have been seven published studies on role of miRNAs and cardiac hypertrophy [Carè *et al.*, 2007; Cheng *et al.*, 2007; Sayed *et al.*, 2007; van Rooij *et al.*, 2006 & 2007; Thum *et al.*, 2007; Tatsuguchi *et al.*, 2007]. The common finding of these studies is that an array of miRNAs is significantly altered in their expression, either up- or down-regulated, and that single miRNAs can critically determine the generation and progression of cardiac hypertrophy. The most consistent changes reported by these studies are up-regulation of *miR-21* (6 of 6 studies), *miR-23a* (4 of 6), *miR-125b* (5 of 6), *miR-214* (4 of 6), *miR-24* (3 of 6), *miR-29* (3 of 6) and *miR-195* (3 of 6), and down-regulation of *miR-1*, *miR-133*, *miR-150* (5 of 6 studies) and *miR-30* (5 of 6). These miRNAs were therefore included in our analysis of target genes encoding ion channel and transporter proteins, as shown in Fig. **4**. Our analyses of these miRNAs gave the following messages.

(1) It is known that cardiac myocytes are characterized with re-expression of the funny current (or pacemaker current) I_f that may underlie the increased risk of arrhythmogenesis in hypertrophic and failing heart [23], which is carried by HCN2 channel in cardiac muscles. We have previously verified that downregulation of *miR-1* and *miR-133* caused upregulation of HCN2 in cardiac hypertrophy [23]. This may contribute to the enhanced abnormal cardiac automaticity and the associated arrhythmias in CHF.

(2) The NCX1 is upregulated in cardiac hypertrophy, ischemia, and failure. This upregulation can have an effect on Ca^{2+} transients and possibly contribute to diastolic dysfunction and an increased risk of arrhythmias [46, 54-58]. Our target prediction indicates that SLC8A1, the gene encoding NXC1 protein, is a potential target for both *miR-1* and *miR-30a/b/c*. The downregulation of *miR-1* and *miR-30a/b/c* in hypertrophy/failure is deemed to relieve the repression of SLC8A1/NCX1 since a strong tonic repression *miR-1* and *miR-30a/b/c* is anticipated considering the high abundance of these miRNAs. On the other hand, upregulation of *miR-214* tends to repress NCX1, but the expression level

of *miR-214* is of no comparison with those of *miR-1* and *miR-30a/b/c*; its offsetting effect should be minimal. Our prediction thus provides a plausible explanation for the upregulation of NCX1 through the miRNA mechanism.

(3) A variety of Na^+ channel abnormalities have been demonstrated in heart failure. Several studies suggest that peak I_{Na} is reduced which can cause slowing of cardiac conduction and promote re-entrant arrhythmias [59-62]. It is been speculated that post-transcriptional reductions of the cardiac I_{Na} □-subunit protein Nav1.5 [62] may account for the reduction of peak I_{Na}. In this study, we found that the only miRNA that can target Nav1.5 and is upregulated in cardiac hypertrophy/CHF is *miR-125a/b*. As an abundantly expressed miRNA, upregulation of *miR-125a/b* could well result in repression of SCN5A/Nav1.5.

(4) The gap junction channel proteins connexin43, connexin45 and connexin40 are important for cell-to-cell propagation of excitations. Downregulation of connexin43 expression is associated with an increased likelihood of ventricular tachyarrhythmias in heart failure [63]. Other connexins, including connexin45 [64] and connexin40 [65], are upregulated in failing hearts, possibly as a compensation for connexin43 downregulation. Our analysis indicates that the upregulation of *miR-125a/b* and *miR-23a/b* should produce repression of connexin43 and connexin45 and the down regulation of *miR-1*, *miR-30a/b/c* and *miR-150* should do the opposite. These two opposing effects may cancel out each other.

(5) Prolongation of ventricular APD is typical of heart failure to enable the improvement of contraction strength and thereby support the weakened heart. However, APD prolongation consequent to decreases in several repolarizing K^+ current (I_{to1}, I_{Ks}, and I_{K1}) in failing heart often results in occurrence of early afterdepolarizations (EAD) [66-71]. Our prediction, however, failed to provide any explanation at the miRNA level: None of the upregulated miRNAs may regulate the genes encoding repolarizing K^+ channels. On the contrary, downregulation of *miR-1* and *miR-133* predict upregulation of KCNE1 and KCNQ1, respectively.

(6) A majority of published studies showed a decrease in I_{K1} in ventricular myocytes of failing hearts [46, 68-71]. But whether KCNJ2/Kir2.1, the major subunit underlying I_{K1}, is downregulated remained controversial in previous studies and the mechanisms remained obscured. One study noted decreased KCNJ2 mRNA expression but unaltered Kir2.1 protein level [71]. With our prediction, the upregulated miRNAs (*miR-125*, *miR-214*, *miR-24*, *miR-29*, and *miR-195*) predict reduction of inward rectifier K^+ channel subunits including KCNJ2/Kir2.1, KCNJ12/Kir2.2, KCNJ14/Kir2.4, and KCNK1/TWIK1, whereas the downregulated miRNAs (*miR-1* and *miR-30a/b/c*) predict increase in KCNJ2/Kir2.1.

In summary, our analysis of target genes for deregulated miRNAs in hypertrophy/CHF may explain at least partly the enhanced cardiac automaticity (relief of HCN2 repression and increased NCX1 expression) and reduced cardiac conduction (repression of Nav1.5). But the data suggest that miRNAs are hardly involved in the abnormality of cardiac repolarization in cardiac hypertrophy and heart failure since the genes for the repolarizing K^+ channels were not predicted as targets for the upregulated miRNAs. The prediction of NCX1 upregulation as a result of derepression from miRNAs may be of particular importance aberrantly enhanced NXC1 activity has also been noticed in atrial fibrillation occurring in CHF.

Myocardial Infarction (MI)

MI, a typical situation of metabolic stress, is presented as cascades of cellular abnormalities as a result of deleterious alterations of gene expression outweighing adaptive changes. MI can cause severe cardiac injuries and the consequences are contraction failure, electrical abnormalities and even lethal arrhythmias, and eventual death of the cell. Ischemic myocardium demonstrates characteristic sequential alterations in electrophysiology with an initial shortening of APD and QT interval during the early phase (<15min) of acute ischemia and subsequent lengthening of APD/QT after a prolonged ischemic period and chronic myocardial ischemia. While these alterations may be adaptive to the altered metabolic status, they occur at the cost of arrhythmogenesis consequent to ischemic ionic remodelling. To exploit if miRNAs could be involved in the remodelling process, several original studies have been published. We first identified upregulation of *miR-1* in acute myocardial infarction and the ischemic arrhythmias caused by this deregulation of *miR-1* expression. Similar ischemic *miR-1* upregulation was reproduced by another two groups. Subsequently, miRNA expression profiles in the setting of myocardial ischemia/reperfusion injuries were reported by three groups. We have also obtained the miRome in a rat model of acute myocardial infarction (Supplementary Fig. S2 in Luo *et al.*, 2010).

Extracting of the overlapping results from different laboratories and filtering with the cardiac expression verified by real-time RT-PCR in human hearts allowed us to identify an array of miRNAs that are likely deregulated in the setting of myocardial ischemia. The upregulated miRNAs include *miR-1, miR-23, miR-29, miR-20, miR-30, miR-146b-5p, miR-193, miR-378, miR-181, miR-491-3p, miR-106, miR-199b-5p,* and *let-7f*; the downregulated miRNAs include *miR-320, miR-185, miR-324-3p,* and *miR-214* (Fig. **4**). This analysis excluded some miRNAs that were found deregulated by a study but not by others and that were found deregulated in rat heart but was not expressed in human heart (e.g. *miR-208* is upregulated in rat model but it is not expressed at all in human heart). Interesting to note is that some of the miRNAs demonstrated the opposite directions of changes in their expression between ischemic myocardium and hypertrophic hearts. For example, *miR-1, let-7, miR-181b, miR-29a* and *miR-30a/e*, which are upregulated in ischemic myocardium, are downregulated in hypertrophy. Similarly, *miR-214, miR-320* and *miR-351*, which are down-regulated in ischemic myocardium, are up-regulated in hypertrophy (Fig. **5**). This fact further reinforces the notion that different pathological conditions have different expression profiles. Our analysis yielded the following notions.

(1)　Six upregulated miRNAs (*miR-1, miR-29, miR-20, miR-30, miR-193* and *miR-181*) were predicted to target several Kir subunits (KCNJ2, KCNJ12, KCNJ, and KCNK1), but none of the downregulated miRNAs can target these genes. This is in line with the previous finding that I_{K1} is reduced and membrane is depolarized in ischemic myocardium [Yang *et al.*, 2007].

(2)　The cardiac slow delayed rectifier K^+ current (I_{Ks}) is carried by coassembly of an α-subunit KvLQT1 (encoded by KCNQ1) and a β-subunit mink (encoded by KCNE1). Loss-of-function mutation of either KCNQ1 or KCNE1 can cause long QT syndromes, indicating the importance of I_{Ks} in cardiac repolarization. In ischemic myocardium, persistent decreases in minK with normalized KvLQT1 protein expression have been observed which may underlie unusual delayed rectifier currents with very rapid activation, resembling currents produced by the expression of KvLQT1 in the absence of minK. We have experimentally established KCNE1 as a target for *miR-1* repression, which was also predicted in the present analysis. Moreover, no other miRNAs were predicted to target KCNQ1. This finding is coincident with the observations on the diminishment of minK alone without changes of KvLQT1 in ischemic myocardium.

(3)　It has been observed that cells in the surviving peri-infarct zone have discontinuous propagation due to abnormal cell-to-cell coupling. This is largely due to decreased expression and redistribution of gap junction protein connexins (Cxs). In this study, seven out of 12 upregulated miRNAs were predicted to target Cxs including GJA1/Cx43, GJC1/Cx45, and GJA5/Cx40, but only one downregulated miRNA *miR-185* may regulate GJA5/Cx40. This result clearly points to the role of miRNAs in damaging cardiac conduction in ischemic myocardium. Indeed, repression of GJA1/Cx43 to slow conduction and induce arrhythmias in acute myocardial infarction has been experimentally verified by our previous study.

(4)　In ischemic myocardium, fast or peak I_{Na} density is reduced, which may also account partly for the conduction slowing and the associated re-entrant arrhythmias [86-88]. Our analysis showed that *let-7f* and *miR-378* may target SCN5A/Nav1.5 and upregulation of these miRNAs is anticipated to cause reduction of I_{Na} via downregulating SCN5A/Nav1.5 in myocardial infarction. By comparison, none of the downregulated miRNAs may repress SCN5A/Nav1.5 based on our target prediction.

(5)　I_{to1} is reduced in myocardial ischemia and in rats, I_{to1} decreases correlate most closely with downregulation of KCND2-encoded Kv4.2 subunits [Gidh-Jain *et al.*, 1996; Nerbonne & Kass, 2005]. *miR-1* is predicted to repress KCND2/Kv4.2, and *miR-29* may target KCHiP2 that is known to be critical in the formation of I_{to1}.

(6)　$I_{Ca,L}$ is diminished in border-zone cells of dogs. *miR-30* has the potential to target CACNA1C/Cav1.2 and CACNB2/Cav□2, and *miR-124, miR-181, miR-320* and *miR-204* to target CACNB2. Upregulation of *miR-30, miR-124* and *miR-181* therefore would decrease CACNA1C/Cav1.2 and CACNB2/Cav□2 expression, but downregulation of *miR-320* and *miR-204* tends to increase the expression of these genes. Considering the relative abundance of these miRNAs, it seems that the decreasing force overweighs the increasing force with a balance towards a net inhibition of $I_{Ca,L}$.

(7)   Na$^+$/K$^+$ ATPase is a sarcolemmal ATP-dependent enzyme transporter that transports three intracellular Na$^+$ ions to the extracellular compartment and moves two extracellular K$^+$ ions into the cell to maintain the physiological Na$^+$ and K$^+$ concentration gradients for generating the rapid upstroke of the action potential but also for driving a number of ion-exchange and transport processes crucial for normal cellular function, ion homeostasis and the control of cell volume. It is electrogenic, producing a small outward current I$_P$. We noticed that the ischemia-induced upregulation of *miR-29* and *miR181* expression might render inhibition of Na$^+$/K$^+$ ATPase activity as they possibly target the ATP1B1 β-subunit of the enzyme. This may contribute to the electrical and contractile dysfunction in the ischemic/reperfused myocardium due to the ischemia-induced inhibition of the Na$^+$/K$^+$ ATPase and the failure of intracellular Na$^+$ to recover completely on reperfusion.

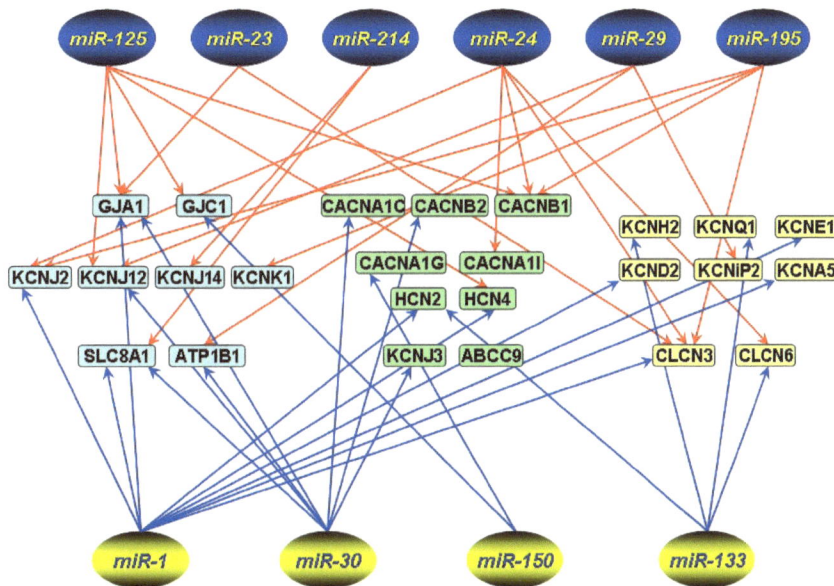

Figure 4: Predicted gene targeting of the miRNAs deregulated in their expression in cardiac hypertrophy and congestive heart failure (CHF). The arrows in red indicate repression of the genes by the upregulated miRNAs (top row in blue) and those in blue indicate derepression of the genes by the downregulated miRNAs (bottom row in yellow). The target genes are roughly divided into groups in three different colors: the genes for cardiac conduction in blue, the genes for cardiac automaticity in green, and the genes for cardiac repolarization in yellow.

In a whole, it appears that the expression signature of miRNAs in the setting of myocardial ischemia and the predicted gene targeting of these miRNAs coincide with the ionic remodelling process under this pathological condition (Fig. **5**). The miRNAs seem to be involved in all aspects of the abnormalities of cardiac excitability during ischemia, as manifested by the slowing of cardiac conduction due to reduced I_{Na} and Cx43, the depolarized membrane potential to adversely affect cardiac conduction due to reduced I_{K1}, the impaired excitation-contraction coupling and contractile function due to reduced $I_{Ca,L}$ and Na$^+$/K$^+$ ATPase, and the delayed cardiac repolarization due to reduced I_{Ks} and I_{to1}.

Atrial Fibrillation (AF)

AF is the most commonly encountered clinical arrhythmia that causes tremendous health problems by increasing the risk of stroke and exacerbating heart failure. It is characterized by a process termed atrial electrical remodeling: the rapid atrial activation rate during AF can remodel the atrial electrophysiology to promote the recurrence and maintenance of AF. A prominent finding in atrial electrical remodeling is shortening of atrial effective refractory period (ERP) favoring re-entrant arrhythmias, primarily because of the shortening of atrial APD as a result of two critical changes. The first change is the reduction of L-type Ca^{2+} current (I_{CaL}) that serves to shorten the plateau duration. And the second change is the increase in inward rectifier K$^+$ current I_{K1}, which underlies the shortening of the terminal phase [96-98]. Whereas it is known that the expression of the genes for these channels is deregulated during AF, the precise molecular mechanisms remained unclear. Additionally, expression of other ion channels such as KCND3 (encoding Kv4.3 for I_{to1}) and KCNA5 (encoding Kv1.5 for I_{Kur}) has been consistently found downregulated in AF, though the role of the changes in AF is yet to be elucidated. The present analysis, however, aids us to get some insight into the issue.

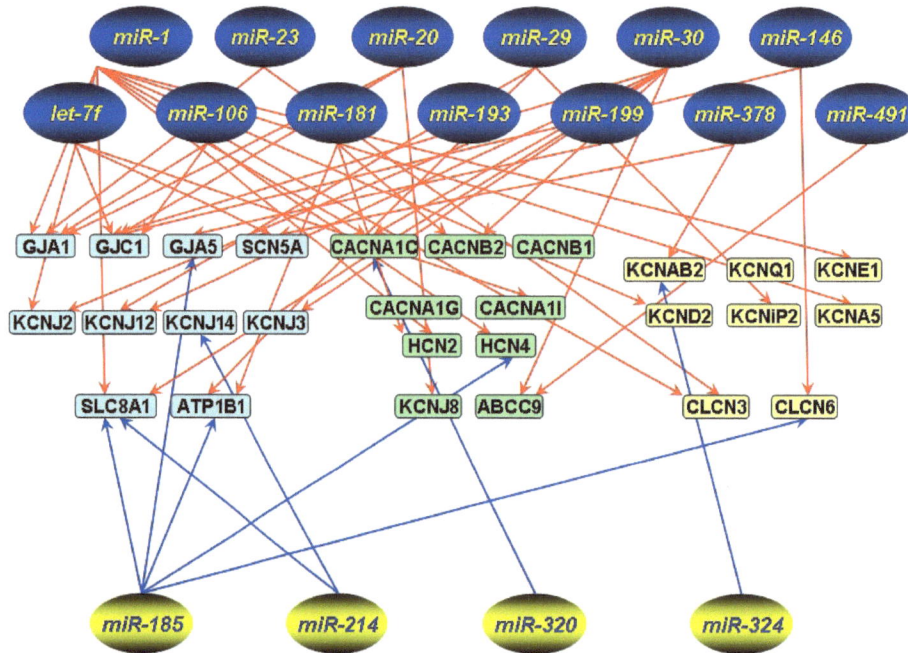

Figure 5: Predicted gene targeting of the miRNAs deregulated in their expression in ischemic myocardial injuries. The arrows in red indicate repression of the genes by the upregulated miRNAs (top rows in blue) and those in blue indicate derepression of the genes by the downregulated miRNAs (bottom row in yellow). The target genes are roughly divided into groups in three different colors: the genes for cardiac conduction in blue, the genes for cardiac automaticity in green, and the genes for cardiac repolarization in yellow.

We first conducted expression profiling to identify deregulated miRNAs in the atrial tissues of a canine model of tachypacing-induce AF, using miRNA microarray analysis comparing the differential expressions of miRNAs between control and AF dogs. Four miRNAs *miR-223, miR-328, miR-664* and *miR-517* were found increased by >2 folds, and six were decreased by at least 50% including *miR-101, miR-133, miR-145, miR-320, miR-373* and *miR-499*. Real-time quantitative RT-PCR (qRT-PCR) analysis confirmed the significant upregulation of *miR-223, miR-328* and *miR-664* (*miR-517* was undetectable), and the significant downregulation of *miR-101, miR-320*, and *miR-499* (Fig. **6**). Our subsequent analysis was therefore based on the deregulated miRNAs verified by qPCR.

(1) Our prediction indicates that three miRNA *miR-328, miR-145* and *miR-320* have the potential to repress both the □1c- and □2-subunits of cardiac L-type Ca^{2+} channel genes, CACNA1C and CACNB2, respectively. While increased miR-328 level should upregulate L-type Ca^{2+} channel expression, decreased *miR-145* and *miR-320* levels should downregulate it. In reality, these two opposing actions may offset each other.

(2) Among the deregulated miRNAs, the only miRNA that may target KCND3 is *miR-328*. Hence, downregulation of *miR-328* predicts downregulation of Kv4.3 thereby reduction of I_{to1} in AF.

(3) Increase in I_{K1} is a hallmark of atrial electrical remodeling in AF. *miR-101* was predicted to target KCNJ2/Kir2.1 and downregulation of this miRNA should upregulate KCNJ2/Kir2.1 due to a relief of repression. Repression of KCNJ12/Kir2.2 due to *miR-328* upregulation may be canceled out by derepression upon *miR-145* downregulation.

(4) Impulse initiation by automaticity and triggered activity as well as impulse initiation resulting from reentry in AF have been suggested. The hyperpolarization activated cation current or funny current I_f is a candidate for contributing to abnormal automaticity. The downregulation of *miR-133* in AF predicts enhancement of I_f through derepression of HCN2 to induce abnormal cardiac automaticity.

(5) Our data did not predict involvement of miRNAs in the alterations of the genes for I_{Kur}, I_{Kr} and I_{Ks}.

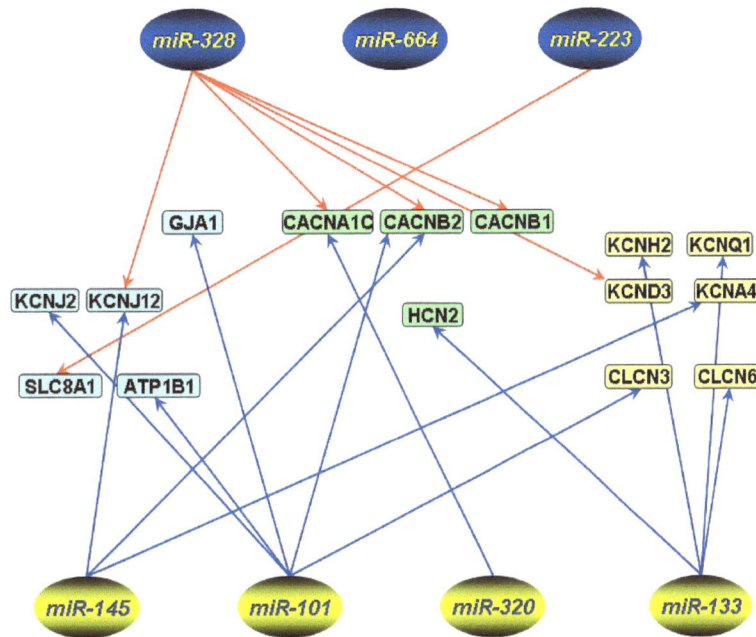

Figure 6: Predicted gene targeting of the miRNAs deregulated in their expression in experimental atrial fibrillation. The arrows in red indicate repression of the genes by the upregulated miRNAs (top row in blue) and those in blue indicate derepression of the genes by the downregulated miRNAs (bottom row in yellow). The target genes are roughly divided into groups in three different colors: the genes for cardiac conduction in blue, the genes for cardiac automaticity in green, and the genes for cardiac repolarization in yellow.

Taken together, it appears that the miRNA expression signature identified in a canine model of tachypacing-induced AF is related to the atrial ionic remodelling process. Specifically, downregulation of *miR-101* and *miR-133* may contribute to enhanced I_{K1} and I_f, respectively, and upregulation of *miR-328* may underlie the reduction of I_{to1}, in AF. Whether this upregulation also contributes to the reduction of $I_{Ca,L}$ need to be examined experimentally. The characteristic decrease I_{Kur} in AF are unlikely related to miRNA deregulation.

Cardiovascular diseases remain the major cause of mortality and morbidity in developed countries. Most of the cardiac deaths are sudden, occurring secondary to ventricular arrhythmias, the electrical disturbances that can result in irregular cardiac contraction. Abnormally altered cardiac excitability suggests that for arrhythmias to arise, the normal matrix of ion channels and transporters must be perturbed by arrhythmogenic substrates to produce a proarrhythmic conditions to permit rhythmic disturbances caused by impaired excitation conduction/propagation, enhanced automaticity, or abnormal repolarization. Cardiac ion channels are fundamental determinants of cardiac excitability. In some cases, abnormalities of these ion channels, channelopathies, can be attributed to mutations in the genes encoding the channel proteins, which can predispose to arrhythmias. In other cases, malfunction of ion channels can also be ascribed to abnormally altered expression. The present study aims to acquire an overall picture about the potential expression regulation of ion channel and transporter genes by miRNAs and the possible implications of this regulatory mechanism. The theoretical analysis in conjunction with experimental demonstration of miRNA expression profiles under various conditions performed in this study allowed us to establish a matrix of miRNAs that are expressed in cardiac cells and have the potential to regulate the genes encoding cardiac ion channels and transporters. These miRNAs likely play an important role in controlling cardiac excitability and keeping the normal electrical activities of the heart. In other words, the ion channel genes may normally be under the post-transcriptional regulation of a group of miRNAs in addition to the muscle-specific miRNAs *miR-1* and *miR-133* as already demonstrated experimentally. Also were we able to link a particular ionic remodeling process in hypertrophy/heart failure, myocardial ischemia, or atrial fibrillation to the corresponding deregulated miRNAs under that pathological condition; the changes of miRNAs appear to have anti-correlation with the changes of many of the genes encoding cardiac ion channels under these situations. These results indicate that multiple miRNAs might be critically involved in the electrical/ionic remodeling processes of cardiac diseases through altering their expression in cardiac cells. Intriguingly, the miRNA targeting under three different conditions clearly demonstrated three different patterns with that in hypertrophy/CHF showing balanced repression and derepression, in MI showing repression overweighing derepression, and in AF showing the opposite: derepression overweighing repression. Another important notion revealed by this study is that though we have elucidated role of *miR-1* and *miR-133* in controlling cardiac excitability and the associated arrhythmogenesis in the above-mentioned three pathological conditions, it is now clearly that other miRNAs that are deregulated are also likely involved in these processes. In reality, it is conceivable that the electrical/ionic remodeling processes under various conditions are caused by many miRNAs in addition to other regulatory molecules.

REFERENCES

Antzelevitch C. (2005) Modulation of transmural repolarization. Ann N Y Acad Sci 1047:314–323.

Babij P, Askew GR, Nieuwenhuijsen B, Su CM, Bridal TR, Jow B, Argentieri TM, Kulik J, DeGennaro LJ, Spinelli W, Colatsky TJ. (1998) Inhibition of cardiac delayed rectifier K_ current by overexpression of the long-QT syndrome HERG G628S mutation in transgenic mice. Circ Res 83:668–678.

Bers DM. (2002) Cardiac excitation–contraction coupling. Nature 415:198–205.

Bosch RF, Zeng X, Grammer JB, Popovic CM, Kuhlkamp V. (1999). Ionic mechanisms of electrical remodelling in human atrial fibrillation. Cardiovasc Res 44:121–131.

Boštjančič E, Zidar N, Stajer D, Glavač D. (2010) MicroRNA miR-1 is up-regulated in remote myocardium in patients with myocardial infarction. Folia Biol (Praha) 56:27–31.

Carè A, Catalucci D, Felicetti F, Bonci D, Addario A, Gallo P, Bang ML, Segnalini P, Gu Y, Dalton ND, Elia L, Latronico MV, Høydal M, Autore C, Russo MA, Dorn GW, Ellingsen O, Ruiz-Lozano P, Peterson KL, Croce CM, Peschle C, Condorelli G. (2007) MicroRNA-133 controls cardiac hypertrophy. Nat Med 13:613–618.

Cha TJ, Ehrlich JR, Zhang L, Nattel S. (2004) Atrial ionic remodelling induced by atrial tachycardia in the presence of congestive heart failure. Circulation 110:1520–1526.

Chen X, Piacentino V 3rd, Furukawa S, Goldman B, Margulies KB, Houser SR. (200) L-type Ca^{2+} channel density and regulation are altered in failing human ventricular myocytes and recover after support with mechanical assist devices. Circ Res 91:517–524.

Cheng Y, Ji R, Yue J, Yang J, Liu X, Chen H, Dean DB, Zhang C. (2007) MicroRNAs are aberrantly expressed in hypertrophic heart. Do they play a role in cardiac hypertrophy? Am J Pathol 170:1831–1840.

Clements-Jewery H, Hearse DJ, Curtis MJ. (2005) Phase 2 ventricular arrhythmias in acute myocardial infarction: a neglected target for therapeutic antiarrhythmic drug development and for safety pharmacology evaluation. Br J Pharmacol 145:551–564.

Costantini DL, Arruda EP, Agarwal P, Kim KH, Zhu Y, Zhu W, Lebel M, Cheng CW, Park CY, Pierce SA, Guerchicoff A, Pollevick GD, Chan TY, Kabir MG, Cheng SH, Husain M, Antzelevitch C, Srivastava D, Gross GJ, Hui CC, Backx PH, Bruneau BG. (2005) The homeodomain transcription factor Irx5 establishes the mouse cardiac ventricular repolarization gradient. Cell 123:347–358.

Davare MA, Horne MC, Hell JW. (2000) Protein phosphatase 2A is associated with class C L-type calcium channels (Cav1.2) and antagonizes channel phosphorylation by cAMP-dependent protein kinase. J Biol Chem 275:39710–39717.

Day CP, McComb JM, Campbell RWF. (1990) QT dispersion: An indication of arrhythmia risk in patients with long QT intervals. Br Heart J 63:342–344.

Gidh-Jain M, Huang B, Jain P, el-Sherif N. (1996) Differential expression of voltage-gated K^+ channel genes in left ventricular remodeled myocardium after experimental myocardial infarction. Circ Res 79:669–675.

Dobrev D, Graf E, Wettwer E, Himmel HM, Hala O, Doerfel C, Christ T, Schuler S, Ravens U. (2001) Molecular basis of downregulation of G-protein-coupled inward rectifying K^+ current $I_{K,ACh}$ in chronic human atrial fibrillation: decrease in GIRK4 mRNA correlates with reduced $I_{K,ACh}$ and muscarinic receptor-mediated shortening of action potentials. Circulation 104:2551–2557.

Dobrev D, Wettwer E, Kortner A, Knaut M, Schuler S, Ravens U. (2002) Human inward rectifier potassium channels in chronic and postoperative atrial fibrillation. Cardiovasc Res 54:397– 404.

Ebihara M, Ohba H, Kikuchi M, Yoshikawa T. (2004) Structural characterization and promoter analysis of human potassium channel Kv8.1 (KCNV1) gene. Gene 325:89–96.

Fernandez-Velasco M, Goren N, Benito G, Blanco-Rivero J, Bosca L, Delgado C. (2003) Regional distribution of hyperpolarization-activated current I_f and hyperpolarization-activated cyclic nucleotide-gated channel mRNA expression in ventricular cells from control and hypertrophied rat hearts. J Physiol 553:395–405.

Gaborit N, Steenman M, Lamirault G, Le Meur N, Le Bouter S, Lande G, Leger J, Charpentier F, Christ T, Dobrev D, Escande D, Nattel S, Demolombe S. (2005) Human atrial ion channel and transporter subunit gene-expression remodeling associated with valvular heart disease and atrial fibrillation. Circulation 112:471–481.

Gintant GA.(1995) Regional differences in I_K density in canine left ventricle: Role of IKs in electrical heterogeneity. Am J Physiol 268:H604–H613.

Girmatsion Z, Biliczki P, Bonauer A, Wimmer-Greinecker G, Scherer M, Moritz A, Bukowska A, Goette A, Nattel S, Hohnloser SH, Ehrlich JR. (2009) Changes in microRNA-1 expression and I_{K1} up-regulation in human atrial fibrillation. Heart Rhythm 6:1802–1809.

Griffiths-Jones S, Saini HK, van Dongen S, Enright AJ. (2008) miRBase: tools for microRNA genomics. Nucleic Acids Res 36:D154–D158.

Grueter CE, Colbran RJ, Anderson ME. (2007) CaMKII, an emerging molecular driver for calcium homeostasis, arrhythmias, and cardiac dysfunction. J Mol Med 85:5–14.

Guo W, Xu H, London B, Nerbonne JM. (1999) Molecular basis of transient outward K^+ current diversity in mouse ventricular myocytes. J Physiol 521:587–599.

Guo W, Li H, London B, Nerbonne JM. (2000) Functional consequences of elimination of $I_{to,f}$ and $I_{to,s}$: early afterdepolarizations, atrioventricular block, and ventricular arrhythmias in mice lacking Kv1.4 and expressing a dominant-negative Kv4 ☐ subunit. Circ Res 87:73–79.

Hall DD, Feekes JA, Arachchige Don AS, Shi M, Hamid J, Chen L, Strack S, Zamponi GW, Horne MC, Hell JW. (2006) Binding of protein phosphatase 2A to the L-type calcium channel Cav1.2 next to Ser1928, its main PKA site, is critical for Ser1928 dephosphorylation. Biochemistry 45:3448–3459.

Idriss SF, Wolf PD. (2004) Transmural action potential repolarization heterogeneity develops postnatally in the rabbit. J Cardiovasc Electrophysiol 15:795–801.

Kaprielian R, Sah R, Nguyen T, Wickenden AD, Backx PH. (2002) Myocardial infarction in rat eliminates regional heterogeneity of AP profiles, I_{to} K^+ currents, and $[Ca^2]_i$ transients. Am J Physiol 283:H1157–H1168.

Lewis BP, Burge CB, Bartel DP. (2005) Conserved seed pairing, often flanked by adenosines, indicates that thousands of human genes are microRNA targets. Cell 120:15–20.

Li GR, Feng J, Yue L, Carrier M. (1998) Transmural heterogeneity of action potentials and I_{to1} in myocytes isolated from the human right ventricle. Am J Physiol 275:H369–H377.

Li Y, Ma J, Jiao JM, Liu N, Niu HY, Lu ZY. (2002) Heterogeniety of action potential and ion currents in the left ventricular myocyte of the rabbit. Acta Physiol 54:369–374.

Liang Y, Ridzon D, Wong L, Chen C. (2007) Characterization of microRNA expression profiles in normal human tissues. BMC Genomics 8: 166.

Liu DW, Antzelevitch C. (1995) Characteristics of the delayed rectifier current (IKr and IKs) in canine ventricular epicardial, midmyocardial, and endocardial myocytes. A weaker IKs contributes to the longer action potential of the M cell. Circ Res 76:351–365.

Liu DW, Gintant GA, Antzelevitch C. (1993) Ionic bases for electrophysiological distinctions among epicardial, midmyocardial, and endocardial myocytes from the free wall of the canine left ventricle. Circ Res 72:671–687.

Liu G, Iden JB, Kovithavongs K, Gulamhusein R, Duff HJ, Kavanagh KM. (2004) *In vivo* temporal and spatial distribution of depolarization and repolarization and the illusive murine T wave. J Physiol 555:267–279.

Lois C, Hong EJ, Pease S, Brown EJ, Baltimore D. (2002) Germline transmission and tissue-specific expression of transgenes delivered by lentiviral vectors. Science 295:868−872.

Lundquist AL, Turner CL, Ballester LY, George Jr AL. (2006) Expression and transcriptional control of human KCNE genes. Genomics 87:119–128.

Luo X, Lin H, Lu Y, Li B, Xiao J, Yang B, Wang Z. (2007) Transcriptional activation by stimulating protein 1 and post-transcriptional repression by muscle-specific microRNAs of I_{Ks}-encoding genes and potential implications in regional heterogeneity of their expressions. J Cell Physiol 212:358–367.

Luo X, Lin H, Xiao J, Zhang Y, Lu Y, Yang B, Wang Z. (2008) Downregulation of *miRNA-1/miRNA-133* contributes to re-expression of pacemaker channel genes *HCN2* and *HCN4* in hypertrophic heart. J Biol Chem 283:20045–20052.

Luo X, Zhang H, Xiao J, Wang Z. (2010) Regulation of human cardiac ion channel genes by microRNAs: Theoretical perspective and pathophysiological implications. *Cell Physiol Biochem* 25:571–586.

Maier LS, Bers DM. (2007) Role of Ca^{2+}/calmodulin-dependent protein kinase (CaMK) in excitation– contraction coupling in the heart. Cardiovasc Res 73:631–640.

Marx SO, Reiken S, Hisamatsu Y, Jayaraman T, Burkhoff D, Rosemblit N, Marks AR. (2000) PKA phosphorylation dissociates FKBP12.6 from the calcium release channel (ryanodine receptor): defective regulation in failing hearts. Cell 101:365–376.

McManus MT, Petersen CP, Haines BB, Chen J, Sharp PA. (2002) Gene silencing using microRNA designed hairpins. RNA 8:842–850.

Matkovich SJ, Wang W, Tu Y, Eschenbacher WH, Dorn LE, Condorelli G, Diwan A, Nerbonne JM, Dorn GW 2nd (2010) MicroRNA-133a protects against myocardial fibrosis and modulates electrical repolarization without affecting hypertrophy in pressure-overloaded adult hearts. Circ Res 106:166–175.

Miranda KC, Huynh T, Tay Y, Ang YS, Tam WL, Thomson AM, Lim B, Rigoutsos I. (2006) A pattern-based method for the identification of microRNA binding sites and their corresponding heteroduplexes. Cell 126:1203–1217.

Nattel S (2002) New ideas about atrial fibrillation 50 years on. Nature 415:219–226.

Nattel S, Maguy A, Le Bouter S, Yeh YH. (2007) Arrhythmogenic ion-channel remodeling in the heart: heart failure, myocardial infarction, and atrial fibrillation. Physiol Rev 87:425–456.

Nerbonne JM, Kass RS. (2005) Molecular physiology of cardiac repolarization. Physiol Rev 85:1205–1253.

Sayed D, Hong C, Chen IY, Lypowy J, Abdellatif M. (2007) MicroRNAs play an essential role in the development of cardiac hypertrophy. Circ Res 100: 416–424.

Shan H, Zhang Y, Lu Y, Zhang Y, Pan Z, Cai B, Wang N, Li X, Feng T, Hong Y, Yang B. (2009) Downregulation of miR-133 and miR-590 contributes to nicotine-induced atrial remodelling in canines. Cardiovasc Res 83:465–472.

Shi Y. (2003) Mammalian RNAi for the masses. Trends Genet 19:9–12.

Sicouri S, Antzelevitch C. (1991) A subpopulation of cells with unique electrophysiological properties in the deep subepicardium of the canine ventricle: The presence of M cell. Circ Res 68:1729–1741.

Silva JM, Li MZ, Chang K, Ge W, Golding MC, Rickles RJ, Siolas D, Hu G, Xu Z. (2005) Second-generation shRNA libraries covering the mouse and human genomes. Nat Genet 37:1281–1288.

Steenaart NAE, Ganim JR, DiSalvo J, Kranias EG. (1992) The phospholamban phosphatase associated with cardiac sarcoplasmic reticulum is a type 1 enzyme. Arch Biochem Biophys 293:17–24.

Stegmeier F, Hu G, Rickles RJ, Hannon GJ, Elledge SJ. (2005) A lentiviral microRNA-based system for single-copy polymerase II-regulated RNA interference in mammalian cells. Proc Natl Acad Sci USA 102:13212–13217.

Stilli D, Sgoifo A, Macchi E, Zaniboni M, De Iasio S, Cerbai E, Mugelli A, Lagrasta C, Olivetti G, Musso E. (2001) Myocardial remodeling and arrhythmogenesis in moderate cardiac hypertrophy in rats. Am J Physiol 280:H142–H150.

Sun D, Melegari M, Sridhar S, Rogler CE, Zhu L. (2006) Multi-miRNA hairpin method that improves gene knockdown efficiency and provides linked multi-gene knockdown. Biotechniques 41:59–63.

Szentadrassy N, Banyasz T, Biro T, Szabo G, Toth BI, Magyar J, Lazar J, Varro A, Kovacs L, Nanasi PP. (2005) Apico-basal inhomogeneity in distribution of ion channels in canine and human ventricular myocardium. Cardiovasc Res 65:851–860.

Tatsuguchi M, Seok HY, Callis TE, Thomson JM, Chen JF, Newman M, Rojas M, Hammond SM, Wang DZ. (2007) Expression of microRNAs is dynamically regulated during cardiomyocyte hypertrophy. J Mol Cell Cardiol 42:1137–1141.

Terentyev D, Belevych AE, Terentyeva R, Martin MM, Malana GE, Kuhn DE, Abdellatif M, Feldman DS, Elton TS, Györke S. (2009) miR-1 overexpression enhances Ca^{2+} release and promotes cardiac arrhythmogenesis by targeting PP2A regulatory subunit B56alpha and causing CaMKII-dependent hyperphosphorylation of RyR2. Circ Res 104:514–521.

Thum T, Galuppo P, Wolf C, Fiedler J, Kneitz S, van Laake LW, Doevendans PA, Mummery CL, Borlak J, Haverich A, Gross C, Engelhardt S, Ertl G, Bauersachs J. (2007) MicroRNAs in the human heart: a clue to fetal gene reprogramming in heart failure. Circulation 116: 258–267.

van Rooij E, Sutherland LB, Qi X, Richardson JA, Hill J, Olson EN. (2007) Control of stress-dependent cardiac growth and gene expression by a microRNA. Science 316: 575–579.

van Rooij E, Sutherland LB, Liu N, Williams AH, McAnally J, Gerard RD, Richardson JA, Olson EN. (2006) A signature pattern of stress-responsive microRNAs that can evoke cardiac hypertrophy and heart failure. Proc Natl Acad Sci USA 103: 18255–18260.

van Wagoner DR, Pond AL, McCarthy PM, Trimmer JS, Nerbonne JM. (1997) Outward K^+ current densities and Kv1.5 expression are reduced in chronic human atrial fibrillation. Circ Res 80:772–781.

Verduyn SC, Vos MA, van der Zande J, van der Hulst FF, Wellens HJ. (1997a) Role of interventricular dispersion of repolarization in acquired torsade-de-pointes arrhythmias: Reversal by magnesium. Cardiovasc Res 34:453–463.

Verduyn SC, Vos MA, van derZande J, KulcsarA, Wellens HJ. (1997b) Further observations to elucidate the role of interventricular dispersion of repolarization and early afterdepolarizations in the genesis of acquired torsade de pointes arrhythmias: A comparison between almokalant and d-sotalol using the dog as its own control. J Am Coll Cardiol 30:1575–1584.

Verduyn SC, Vos MA, van der Zande J, van der Hulst FF, Wellens HJ. (1997) Role of interventricular dispersion of repolarization in acquired torsade-de-pointes arrhythmias: reversal by magnesium. Cardiovasc Res 34:453–463.

Verduyn SC, Vos MA, van der Zande J, Kulcsar A, Wellens HJ. (1997) Further observations to elucidate the role of interventricular dispersion of repolarization and early afterdepolarizations in the genesis of acquired torsade de pointes arrhythmias: a comparison between almokalant and d-sotalol using the dog as its own control. J Am Coll Cardiol 30:1575–1584.

Volders PG, Sipido KR, Carmeliet E, Spatjens RL, Wellens HJ, Vos MA. (1999) Repolarizing K$^+$ currents I$_{TO1}$ and I$_{Ks}$ are larger in right than left canine ventricular midmyocardium. Circulation 99:206–210.

Volders PG, Sipido KR, Vos MA, Spatjens RL, Leunissen JD, Carmeliet E, Wellens HJ. (1999b) Downregulation of delayed rectifier Kt currents in dogs with chronic complete atrioventricular block and acquired torsades de pointes. Circulation 100:2455–2461.

Wehrens XH, Marks AR. (2004) Novel therapeutic approaches for heart failure by normalizing calcium cycling. Nat Rev Drug Discov 3:565–573.

Wettwer E, Amos GJ, Posival H, Ravens U. (1994) Transient outward current in human ventricular myocytes of subepicardial and subendocardial origin. Circ Res 75:473–482.

Workman AJ, Kane KA, Rankin AC. (2001) The contribution of ionic currents to changes in refractoriness of human atrial myocytes associated with chronic atrial fibrillation. Cardiovasc Res 52:226–235.

Xia XG, Zhou H, Xu ZS. (2005) Promises and challenges in developing RNAi as a research tool and therapy for neurodegenerative diseases. Neurodegenerative Diseases 2:220–231.

Xia XG, Zhou H, Xu Z. (2006a) Multiple shRNAs expressed by an inducible pol II promoter can knock down the expression of multiple target genes. Biotechniques 41:64–68.

Xia XG, Zhou H, Samper E, Melov S, Xu Z. (2006b) Pol II-expressed shRNA knocks down Sod2 gene expression and causes phenotypes of the gene knockout in mice. PLoS Genetics 2:e10.

Xiao F, Zuo Z, Cai G, Kang S, Gao X, Li T. (2009) miRecords: an integrated resource for microRNA-target interactions. Nucleic Acids Res 37:D105–D110.

Xiao JF, Fischer C, Steinlein OK. (2001) Cloning and mutation analysis of the human potassium channel KCNQ2 gene promoter. NeuroReport 12:3733–3739.

Xiao J, Luo X, Lin H, Zhang Y, Lu Y, Wang N, Zhang YQ, Yang B, Wang Z. (2007) MicroRNA *miR-133* represses HERG K$^+$ channel expression contributing to QT prolongation in diabetic hearts. J Biol Chem 282:12363–12367.

Xiao J, Yang B, Lin H, Lu Y, Luo X, Wang Z. (2007) Novel approaches for gene-specific interference via manipulating actions of microRNAs: examination on the pacemaker channel genes HCN2 and HCN4. J Cell Physiol 212:285–292.

Xiao L, Xiao J, Luo X, Lin H, Wang Z, Nattel S. (2008) Feedback remodeling of cardiac potassium current expression: a novel potential mechanism for control of repolarization reserve. Circulation 118:983–992.

Yang B, Lin H, Xiao J, Lu Y, Luo X, Li B, Zhang Y, Xu C, Bai Y, Wang H, Chen G, Wang Z. (2007) The muscle-specific microRNA miR-1 causes cardiac arrhythmias by targeting GJA1 and KCNJ2 genes. Nat Med 13:486–491.

Zeng Y, Cullen BR. (2005a) Efficient processing of primary microRNA hairpins by Drosha requires flanking nonstructured RNA sequences. J Biol Chem 280:27595–27603.

Zeng Y, Cai X, Cullen BR. (2005b) Use of RNA polymerase II to transcribe artificial microRNAs. Methods Enzymol 392:371–380.

Zeng Y, Wagner EJ, Cullen BR. (2002) Both natural and designed micro RNAs can inhibit the expression of cognate mRNAs when expressed in human cells. Mol Cell 9:1327–1333.

Zhang Y, Xiao J, Wang H, Luo X, Wang J, Villeneuve LR, Zhang H, Bai Y, Yang B, Wang Z. (2006) Restoring depressed HERG K$^+$ channel function as a mechanism for insulin treatment of the abnormal QT prolongation and the associated arrhythmias in diabetic rabbits. Am J Physiol 291:1446–1455.

Zhao Y, Ransom JF, Li A, Vedantham V, von Drehle M, Muth AN, Tsuchihashi T, McManus MT, Schwartz RJ, Srivastava D. (2007) Dysregulation of cardiogenesis, cardiac conduction, and cell cycle in mice lacking miRNA-1-2. Cell 129:303–317.

Zhou H, Xia XG, Xu Z. (2005) An RNA polymerase II construct synthesizes short-hairpin RNA with a quantitative indicator and mediates highly efficient RNAi. Nucleic Acids Res 33:e62.

Zhou J, Kodirov S, Murata M, Buckett PD, Nerbonne JM, Koren G. (2003) Regional upregulation of Kv2.1-encoded current, $I_{K,slow2}$, in Kv1DN mice is abolished by crossbreeding with Kv2DN mice. Am J Physiol 284:H491–H500.

CHAPTER 7

miRNAs in Vascular Angiogenesis

Abstract: Several processes including endothelial angiogenesis, vascular neointimal lesion formation, vascular inflammation process, lipoprotein metabolism, and hypertension are critically involved in antherogenesis or atherosclerosis. This chapter aims to introduce the role of miRNAs in endothelial angiogenesis. Angiogenesis is the growth of new blood vessels from pre-existing vessels. Functional endothelial cells are required for angiogenesis, a process involving proliferation, migration, and maturation of endothelial cells. Several miRNAs have been shown to critically regulate vascular angiogenesis and they can be classified into pro-angiogenic miRNAs and anti-angiogenic miRNAs. Pro-angiogenic miRNAs indentified to date include miR-130a, miR-17~92 cluster, miR-210, miR-378, miR-7f, miR-27b, and miR-126. The known anti-angiogenic miRNAs include miR-320, miR-221 and miR-222. In addition, the miRNAs that are reportedly involved in tumor vascular biology are also introduced in this chapter.

INTRODUCTION

Angiogenesis, the growth of new blood vessels from pre-existing vessels, is an important physiological and pathological process, both in normal development and in diseases such as ischaemic cardiovascular diseases and cancers [Tirziu & Simons, 2005; Tortora *et al.*, 2004]. Functional endothelial cells are required for angiogenesis, the growth of new blood vessels during neovascularization from pre-existing vessels, a process involving proliferation, migration, and maturation of endothelial cells [Urbich *et al.*, 2008]. The endothelium also acts as a barrier and serves as the primary sensor of blood flow-mediated mechanotransduction. The integrity of the endothelial monolayer maintained by the coordinated regulation of angiogenesis is therefore fundamental for the homoeostasis of the vascular system. Improvement of endothelial cell function, enhancement of angiogenesis after critical ischaemia, and inhibition of angiogenesis during tumour growth have been considered rational therapeutic strategies.

Under pathological settings, increased myocardial mass might not be accompanied by the appropriate increase in vascularization that is associated with normal growth patterns [Hudlicka *et al.*, 1992]. Coronary angiogenesis, which controls capillary density, is, therefore, an important process determining whether or not the myocardium becomes hypoxic, a state that leads to contractile dysfunction and cardiomyocyte loss. Adequate myocardial perfusion is essential for maintaining cardiac myocyte viability, particularly in the failing heart, wherein LV filling pressures are elevated and subendocardial perfusion is compromised. Prior studies have suggested that fewer capillaries are present in end-stage dilated cardiomyopathy and postpartum cardiomyopathy. The failure to appropriately regulate angiogenesis and vasculogenesis in the failing heart can result in myocardial ischemia with subsequent loss of cardiac myocytes and/or interstitial fibrosis and may thus contribute to progressive myocardial dysfunction in heart failure.

Although multiple growth factors have been shown to regulate angiogenesis and vascular development, recent work suggests that miRNAs act as a regulator at a higher level upstream the protein regulators. There is increasing evidence that miRNAs may regulate vascular integrity, angiogenesis, and wound repair. Indeed, a number of pro- and antiangiogenic miRNAs have been identified that target key proteins involved in endothelial tube formation, as well as endothelial cell proliferation, migration, and apoptosis. Deciphering the miRNA network responsible for the fine-tuning of the angiogenic process might lead to new therapeutic approaches to modulate angiogenesis, potentially useful in ischemic conditions such as myocardial ischemia, peripheral vascular disease and vascular diabetic complications. However, it should be emphasized that many of the miRNAs that have been studied thus far have been identified in neoplastic tissue and may therefore not be relevant to cardiovascular physiology.

The first series of observations establishing the key significance of miRNAs in the regulation of mammalian vascular biology came from experimental studies involved in arresting miRNA biogenesis to deplete the miRNA pools of vascular tissues and cells [Yang *et al.*, 2005; Kuehbacher *et al.*, 2007; Suarez *et al.*, 2007; Shilo *et al.*, 2008]. Homozygous mutant mice were not viable, therefore the embryos were examined. Starting from embryonic day 11.5, virtually all Dicer-null embryos were growth and developmentally retarded compared with their wild-type or heterozygous littermates. The embryos that were still viable at this stage had thin and suboptimally developed blood vessels, providing the very first evidence for the involvement of miRNAs in angiogenesis [Yang *et al.*, 2005].

Microscopic examination of the yolk sacs from the mutant embryos revealed that there were fewer blood vessels in the Dicer-null yolk sacs and that these vessels were thin, small and less organized than those of control yolk sacs. These observations indicate that Dicer-dependent biogenesis of miRNA is required for blood vessel development during embryogenesis.

Subsequent evidence for the involvement of miRNAs in angiogenesis came from *in vitro* studies [Poliseno *et al.*, 2006]. Large-scale analysis of miRNA expression in human umbilical vein endothelial cells (HUVEC) led to the observation that among those miRNAs that are highly expressed, there are 15 that may target the receptors of angiogenic factors. In particular, miR-221 and miR-222 were identified as regulating c-Kit expression as well as the angiogenic properties of the c-kit ligand: stem cell factor. The miR-221/miR-222 and c-Kit interaction represents an integral component of a complex circuit that controls the ability of endothelial cells to form new capillaries [Poliseno *et al.*, 2006]. c-kit is involved in neovascularization [Strumberg, 2005; Roboz *et al.*, 2006] and inhibition of c-kit results in reduced vascular endothelial growth factor (VEGF) expression [Litz & Krystal, 2006].

A study by Suárez *et al* [2007 & 2008] showed that knockdown of Dicer by RNAi techniques in human endothelial cells (ECs) altered the expression of several key regulators of endothelial biology and angiogenesis, including TEK/Tie-2, KDR/VEGFR2, Tie-1, endothelial nitric oxide synthase (eNOS) and IL-8. The RNAi-mediated knockdown of Dicer in ECs, resulted in a significant reduction, but not complete loss of mature miRNAs. The net phenotype of Dicer knockdown in ECs was an increase in eNOS protein levels and NO release, and decrease in EC growth. Forced expression of miR-222 and miR-221 that were among the highest expressed in ECs regulates eNOS protein levels after Dicer silencing.

Dimmeler's laboratory also investigated the role of Dicer and Drosha in angiogenesis [Kuehbacher *et al.*, 2007]. In their study, ECs were transfected with siRNA against Dicer and Drosha to inhibit miRNA biogenesis. Genetic silencing of Dicer and Drosha significantly reduced capillary sprouting of ECs and tube forming activity. Migration of ECs was significantly decreased in Dicer siRNA–transfected cells, whereas Drosha siRNA had no effect. Silencing of Dicer but not of Drosha reduced angiogenesis in vivo. They then screened 168 human miRNAs using real-time PCR and revealed that members of the let-7 family, miR-21, miR-126, miR-221, and miR-222 are highly expressed in endothelial cells. Dicer and Drosha siRNA reduced lef-7f and miR-27b expression. Inhibitors against let-7f and miR-27b also reduced sprout formation indicating that these miRNAs promote angiogenesis by targeting antiangiogenic genes. In silico analysis of predicted targets for let-7 cluster identified the endogenous angiogenesis inhibitor thrombospondin-1. They confirmed that Dicer and Drosha siRNA significantly increased the expression of thrombospondin-1. Their findings indicate that inhibition of Dicer impairs angiogenesis *in vitro* and in vivo, whereas inhibition of Drosha induces only a minor antiangiogenic effect.

These above studies laid the groundwork for later in-depth investigations into the role of individual miRNAs in regulating vascular angiogenesis. The available data suggest that the miRNAs involved in vascular angiogenesis could be generally classified into two categories: the proangiogenic miRNAs and antiangiogenic miRNAs.

PROANGIOGENIC miRNAs

miR-130a

miR-130a is expressed at low levels in quiescent HUVEC and is upregulated in response to foetal bovine serum [Chen & Gorski, 2008]. miR-130a downregulates the anti-angiogenic homeobox proteins GAX (growth arrest homeobox) and HoxA5, and functionally antagonized the inhibitory effects of GAX on endothelial cells proliferation, migration and tube formation and the inhibitory effects of HoxA5 on tube formation vitro [Chen & Gorski, 2008]. Thus, expression of miR-130a antagonized the inhibitory effects of GAX or HOXA5 on endothelial cell tube formation *in vitro*.

miR-17~92 cluster

miR-17~92-transduced tumour cells formed larger, better-perfused tumours *in vivo* [Dews *et al.*, 2006]. The enhanced neovascularization correlated with downregulation of anti-angiogenic thrombospondin-1 (Tsp1) and related proteins, such as connective tissue growth factor (CTGF). Both Tsp1 and CTGF are predicted targets for repression by the miR-17~92 cluster, which was upregulated in colonocytes coexpressing K-Ras and c-Myc. miR-17~92 knockdown with antisense 2'-O-methyl oligoribonucleotides partly restored Tsp1 and CTGF expression; in addition, transduction of Ras-only cells with a miR-17~92-encoding retrovirus reduced Tsp1 and CTGF levels [Dews *et al.*, 2006].

The miR-17~92 cluster is highly expressed in human endothelial cells and that miR-92a, a component of this cluster, controls the growth of new blood vessels (angiogenesis). Forced overexpression of miR-92a in endothelial cells blocked angiogenesis *in vitro* and *in vivo*. In mouse models of limb ischemia and myocardial infarction, systemic administration of an antagomir designed to inhibit miR-92a led to enhanced blood vessel growth and functional recovery of damaged tissue. miR-92a appears to target mRNAs corresponding to several proangiogenic proteins, including the integrin subunit alpha5. Thus, miR-92a may serve as a valuable therapeutic target in the setting of ischemic disease [Bonauer *et al.*, 2009].

miR-210

The expression of miR-210 progressively increased upon exposure to hypoxia [Fasanaro *et al.*, 2008]. miR-210 overexpression in normoxic endothelial cells stimulated the formation of capillary-like structures on Matrigel and vascular endothelial growth factor-driven cell migration. Conversely, miR-210 blockade via anti-miRNA transfection inhibited the formation of capillary-like structures stimulated by hypoxia and decreased cell migration in response to vascular endothelial growth factor. miR-210 overexpression did not affect endothelial cell growth in both normoxia and hypoxia. However, anti-miR-210 transfection inhibited cell growth and induced apoptosis, in both normoxia and hypoxia. Ephrin-A3 was identified as a relevant target of miR-210 in hypoxia since miR-210 was necessary and sufficient to down-modulate its expression. Moreover, luciferase reporter assays showed that Ephrin-A3 was a direct target of miR-210. Ephrin-A3 modulation by miR-210 had significant functional consequences; indeed, the expression of an Ephrin-A3 allele that is not targeted by miR-210 prevented miR-210-mediated stimulation of both tubulogenesis and chemotaxis. Thus, miR-210 up-regulation is a crucial element of endothelial cell response to hypoxia, affecting cell survival, migration, and differentiation [Fasanaro *et al.*, 2008].

miR-378

Expression of miR-378 was shown to enhance cell survival, reduces caspase-3 activity, and promotes tumor growth and angiogenesis [Lee *et al.*, 2007]. Proteomic analysis indicates reduced expression of suppressor of fused (Sufu), a potential target of miR-378, which was confirmed *in vitro* and in vivo. Expression of a luciferase construct containing the target site in Sufu was repressed when cotransfected with miR-378. Transfection of a Sufu construct reversed the effect of miR-378, suggesting an important role for miR-378 in tumor cell survival. miR-378 targets Fus-1. Expression of luciferase constructs harboring the target sites in Fus-1 was repressed by miR-378. Fus-1 constructs with or without its 3' UTR were also generated. Cotransfection experiments showed that the presence of miR-378 repressed Fus-1 expression. Suppression of Fus-1 expression by siRNA against Fus-1 enhanced cell survival. Transfection of the Fus-1 construct reversed the function of miR-378 in cell survival. The results suggest that miR-378 transfection enhanced cell survival, tumor growth, and angiogenesis through repression of the expression of two tumor suppressors, Sufu and Fus-1.

let-7f and miR-27b

A screening analysis of 168 human miRNAs using real-time PCR revealed that members of the let-7 family, miR-21, miR-126, miR-221, and miR-222 are highly expressed in endothelial cells [Kuehbacher *et al.*, 2007]. Dicer and Drosha siRNA reduced lef-7f and mir-27b expression. Inhibitors against let-7f and mir-27b also reduced sprout formation indicating that let-7f and mir-27b promote angiogenesis by targeting antiangiogenic genes. *In silico* analysis of predicted targets for let-7 cluster identified the endogenous angiogenesis inhibitor thrombospondin-1. Indeed, Dicer and Drosha siRNA significantly increased the expression of thrombospondin-1.

miR-126

miR-126 is specifically and highly expressed in human endothelial cells (ECs) and has been shown to regulate the response of ECs to VEGF, and to be important for neovascularization after myocardial infarction through the inhibition of negative regulators of the VEGF pathway, such as SPRED1 and PIK3R2 [Fish *et al.*,. 2008; Wang *et al.*, 2008]. Targeted deletion of the endothelial cell-restricted miRNA miR-126 causes leaking of vessels, haemorrhaging and partial embryonic lethality, due to a loss of vascular integrity and defects in endothelial cell proliferation, migration, and angiogenesis. Mutant surviving mice display defective cardiac neovascularization following myocardial infarction. A link was demonstrated between miR-126 and VEGF and FGF in that it increases the pro-angiogenic effects of these two cytokines [Fish *et al.*, 2008; Wang *et al.*, 2008].

Srivastava and colleagues [Fish *et al.*, 2008] found that miR-126 was enriched in endothelial cells derived from mouse embryonic stem cells and in developing mouse embryos. Subsequent studies in zebrafish showed that miR-126 knockdown resulted in loss of vascular integrity, as well as hemorrhage during embryonic development. These authors found that miR-126 directly repressed 2 negative regulators of the vascular endothelial growth factor (VEGF) pathway, namely Sprouty-related protein (Spred1) and the phosphoinositol-3 kinase regulatory subunit 2 (PIK3R2/p85-α). Increased expression of Spred1 or inhibition of VEGF signaling in zebrafish resulted in defects similar to miR-126 knockdown. Analogous studies in mice revealed that endothelial cell-restricted expression of miR-126–mediated developmental angiogenesis, whereas targeted deletion of miR-126 resulted in leaky vessels, hemorrhaging, and partial embryonic lethality caused by a loss of vascular integrity and defects in endothelial cell proliferation, migration, and angiogenesis. Relevant to the present discussion, the subset of mutant mice that were not embryonic lethal had impaired myocardial neovascularization following myocardial infarction. It was further shown that miR-126 enhanced the proangiogenic actions of VEGF and fibroblast growth factor and promoted formation of new blood vessels by repressing the expression of Spred1.31 Taken together, these findings illustrate that a miR-126 is sufficient to regulate vascular integrity and angiogenesis. However, the relevance of these experimental findings with respect to human heart failure is unclear, insofar as miR-126 is upregulated in failing human hearts [Fish *et al.*,. 2008].

ANTIANGIOGENIC miRNAs

miR-320

One study reported that a number of miRNAs with predicted angiogenicfactor targets are upregulated in myocardial endothelial cells in rats with type 2 diabetes and lead to impaired angiogensis: one of these miRNAs, miR-320, was found to target several angiogenic factors, including Flk1, VEGFc, IGF-1, IGF-1R and FGF, and thus be important for angiopathy [Urbich *et al.*, 2008].

miR-221 and miR-222

The effect of miR-221 and miR-222 on vascular endothelial cell migration was initially determined by tube formation and wound healing assays [Poliseno *et al.*, 2006]. miR-221 and miR-222 inhibit endothelial cell migration, proliferation and angiogenesis by targeting the stem cell factor receptor, ckit, and indirectly regulating eNOs expression [Urbich *et al.*, 2008].

During the development of diabetes, hyperglycemia may cause malfunction of the vasculature. Persistent hyperglycemia causes endothelial cell dysfunction. Li *et al* [2009] observed that exposure to high levels of glucose, which mimics hyperglycemia, induced expression of miR-221 but reduced expression of c-kit, the receptor for stem cell factor in human umbilical vein endothelial cells (HUVECs), which plays a key role in endothelial progenitor cell migration and homing [Poliseno *et al.*, 2006]. Meanwhile, high glucose treatment impaired endothelial cell migration. Incubation with the antisense miR-221 oligonucleotide AMO-221 reduced expression of miR-221 and restored c-kit protein expression in HUVECs treated with high levels of glucose. Furthermore, AMO-221 treatment abolished the inhibitory effect of high glucose exposure on HUVECs transmigration. It appears therefore that miR-221 is induced in HUVECs, which consequently triggers inhibition of c-kit and impairment of HUVECs migration, under hyperglycemic conditions. These findings suggest that manipulation of the miR-221-c-kit pathway may offer a novel strategy for treatment of vascular dysfunction in diabetic patients [Li *et al.*, 2009].

TUMOR VASCULAR BIOLOGY

miR-17~92 Cluster

Human adenocarcinomas commonly harbor mutations in the K-Ras and Myc proto-oncogenes and the TP53 tumor suppressor gene. All three genetic lesions are potentially pro-angiogenic, as they sustain VEGF production.

Interestingly, Kras-transformed mouse colonocytes lacking p53 form indolent, poorly vascularized tumors, whereas additional transduction with a Myc-encoding retrovirus promote vigorous vascularization and growth [Dews *et al.*, 2006]. In addition, VEGF levels were unaffected by Myc, but enhanced neovascularization correlated with down-regulation of anti-angiogenic thrombospondin-1 (Tsp1) and related proteins, such as connective tissue growth factor (CTGF). Both Tsp1 and CTGF are predicted targets for repression by the miR-17~92 miRNA cluster, which is up-regulated in colonocytes co-expressing K-Ras and c-Myc. miR-17~92 knockdown partly restores Tsp1 and CTGF expression. Furthermore, transduction of Ras-only cells with a miR-17~92-encoding retrovirus attenuated Tsp1 and CTGF levels. Cells transduced with miR-17~92, also known as Oncomir-1, formed larger, better-perfused tumors. These findings establish a role for the miR-17~92 cluster in non-cell-autonomous Myc-induced tumor vascular biology [Dews *et al.*, 2006; Ji *et al.*, 2007; Mendell, 2008]. More recent work demonstrates that VEGF regulates the expression of several miRNAs, including cluster miR-17~92. Transfection of endothelial cells with components of the miR-17~92 cluster, induced by VEGF treatment, rescued the induced expression of Tsp1 and the defect in endothelial cell proliferation and morphogenesis initiated by the loss of Dicer [Suárez *et al.*, 2008]. The miR-17~92 cluster composed of seven mature miRNAs (miR-17–5p, miR-17–3p, miR-18a, miR-19a, miR-20a, miR-19b and miR-92–1), residing in intron 3 of the C13orf25 gene at 13q31.3, is markedly and frequently overexpressed in lung cancers, with occasional gene amplification, especially in those with small-cell lung cancer histology [Hayashita *et al.*, 2005].

miR-27a

Genomic estrogen receptor (ER) associates with other transcription factors such as the activating protein-1 complex, nuclear factor-κB and specificity proteins (Sp) to modulate ligand-dependent gene expression. Members of the Sp family of transcription factors exert differential effects on gene transcription. For example, Sp1 and Sp4 act as transcriptional activators, whereas Sp3 antagonizes Sp1 activation by competing for promoter occupancy. Sp1, Sp3 and Sp4 are overexpressed in tumors and contribute to the proliferative and angiogenic phenotype associated with cancer cells [Mertens-Talcott *et al.*, 2007]. Sp1, Sp3 and Sp4 are expressed in a panel of ER-positive and ER-negative breast cancer cell lines and are subject to the regulation by miR-27a, which is also expressed in these cell lines and has been reported to regulate the zinc finger ZBTB10 gene, a putative Sp repressor. Experimental down-regulation of miR-27a using an anti-sense approach increased the expression of ZBTB10 mRNA and decreased the expression of Sp1, Sp3 and Sp4 at the mRNA and protein level, and also decreased activity in cells transfected with constructs containing Sp1 and Sp3 promoter inserts. Such responses were accompanied by decreased expression of Sp-dependent survival and angiogenic genes, including survivin, VEGF, and VEGF receptor 1 [Mertens-Talcott *et al.*, 2007]. Thus, miR-27a supports angiogenesis in the context of breast cancer.

miR-378

In breast cancer tissue, miRNAs regulate tumor vascular invasion [Iorio *et al.*, 2005]. Intravasation of tumor cells into the vasculature and/or lymphatics represents a key stage in cancer metastases. miR-378 enhances cell survival, reduces caspase-3 activity, and promotes tumor growth and angiogenesis by attenuating the expression of suppressor of fused (Sufu), a potential target of miR-378 and Fus-1 [Lee *et al.*, 2007].

REFERENCES

Boettger T, Beetz N, Kostin S, Schneider J, Krüger M, Hein L, Braun T. (2009) Acquisition of the contractile phenotype by murine arterial smooth muscle cells depends on the Mir143/145 gene cluster. J Clin Invest 119:2634–2647.

Bonauer A, Carmona G, Iwasaki M, Mione M, Koyanagi M, Fischer A, Burchfield J, Fox H, Doebele C, Ohtani K, Chavakis E, Potente M, Tjwa M, Urbich C, Zeiher AM, Dimmeler S. (2009) MicroRNA-92a controls angiogenesis and functional recovery of ischemic tissues in mice. Science 324:1710–1713.

Brummelkamp TR, Bernards R, Agami R. (2002) A system for stable expression of short interfering RNAs in mammalian cells. Science 296:550–553.

Chen Y, Chen H, Hoffmann A, Cool DR, Diz DI, Chappell MC, Chen A, Morris M. (2006) Adenovirus-mediated small-interference RNA for *in vivo* silencing of angiotensin AT1a receptors in mouse brain. Hypertension 47:230–237.

Chen Y, Gorski DH. (2008) Regulation of angiogenesis through a microRNA (miR-130a) that down-regulates antiangiogenic homeobox genes GAX and HOXA5. Blood 111:1217–1226.

Chen Y, Leal AD, Patel S, Gorski DH. (2007) The homeobox gene GAX activates p21WAF1/CIP1 expression in vascular endothelial cells through direct interaction with upstream AT-rich sequences. J Biol Chem 282:507–517.

Cheng Y, Liu X, Yang J, Lin Y, Xu DZ, Lu Q, Deitch EA, Huo Y, Delphin ES, Zhang C. (2009) MicroRNA-145, a novel smooth muscle cell phenotypic marker and modulator, controls vascular neointimal lesion formation. Circ Res 105:158–166

Choi WY, Giraldez AJ, Schier AF. (2007) Target protectors reveal dampening and balancing of Nodal agonist and antagonist by miR-430. Science 318:271–274.

Cordes KR, Sheehy NT, White MP, Berry EC, Morton SU, Muth AN, Lee TH, Miano JM, Ivey KN, Srivastava D. (2009) miR-145 and miR-143 regulate smooth muscle cell fate and plasticity. Nature 460:705–710.

Coumoul X, Li W, Wang RH, Deng C. (2004) Inducible suppression of Fgfr2 and Survivin in ES cells using a combination of the RNA interference (RNAi) and the Cre-LoxP system. Nucleic Acids Res 32:e85.

Costinean S, Zanesi N, Pekarsky Y, Tili E, Volinia S, Heerema N, Croce CM. (2006) Pre-B cell proliferation and lymphoblastic leukemia/high-grade lymphoma in E(mu)-miR155 transgenic mice. Proc Natl Acad Sci USA 103:7024–7029.

Dalmay T, Edwards DR. (2006) MicroRNAs and the hallmarks of cancer. Oncogene 25:6170–6175.

Dews M, Homayouni A, Yu D, Murphy D, Sevignani C, Wentzel E, Furth EE, Lee WM, Enders GH, Mendell JT, Thomas-Tikhonenko A. (2006) Augmentation of tumor angiogenesis by a Myc-activated microRNA cluster. Nat Genet 38:1060–1065.

Elia L, Quintavalle M, Zhang J, Contu R, Cossu L, Latronico MVG, Peterson KL, Indolfi C, Catalucci D, Chen J, Courtneidge SA, Condorelli G. (2009) The knockout of miR-143 and -145 alters smooth muscle cell maintenance and vascular homeostasis in mice: correlates with human disease. Cell Death Diff 16:1590–1598.

Everett CA, West JD. (1996) The influence of ploidy on the distribution of cells in chimaeric mouse blastocysts. Zygote 4:59–66.

Fairbrother WG, Yeh RF, Sharp PA, Burge CB. (2002) Predictive identification of exonic splicing enhancers in human genes. Science 297:1007–1013.

Fasanaro P, D'Alessandra Y, Di Stefano V, Melchionna R, Romani S, Pompilio G, Capogrossi MC, Martelli F. (2008) MicroRNA-210 modulates endothelial cell response to hypoxia and inhibits the receptor tyrosine-kinase ligand Ephrin-A3. J Biol Chem 283:15878–15883.

Fish JE, Santoro MM, Morton SU, Yu S, Yeh RF, Wythe JD, Ivey KN, Bruneau BG, Stainier DY, Srivastava D. (2008) miR-126 regulates angiogenic signaling and vascular integrity. Dev Cell 15:272–284.

Fraser AG, Kamath RS, Zipperlen P, Martinez-Campos M, Sohrmann M, Ahringer J. (2000) Functional genomic analysis of C. elegans chromosome I by systematic RNA interference. Nature 408:325–330.

Gao X, Zhang P. (2007) Transgenic RNA Interference in Mice. Physiology 22:161-166.

Hayashita Y, Osada H, Tatematsu Y, Yamada H, Yanagisawa K, Tomida S, Yatabe Y, Kawahara K, Sekido Y, Takahashi T. (2005) A polycistronic microRNA cluster, miR-17-92, is overexpressed in human lung cancers and enhances cell proliferation. Cancer Res 65:9628–9632.

Hudlicka O, Brown M, Egginton S. (1992). Angiogenesis in skeletal and cardiac muscle. Physiol Rev 72:369–417.

Iorio MV, Ferracin M, Liu CG, Veronese A, Spizzo R, Sabbioni S, Magri E, Pedriali M, Fabbri M, Campiglio M, Menard S, Palazzo JP, Rosenberg A, Musiani P, Volinia S, Nenci I, Calin GA, Querzoli P, Negrini M, Croce CM. (2005) MicroRNA gene expression deregulation in human breast cancer. Cancer Res 65:7065–7070.

Ji R, Cheng Y, Yue J, Yang J, Liu X, Chen H, Dean DB, Zhang C. (2007) MicroRNA expression signature and antisense-mediated depletion reveal an essential role of MicroRNA in vascular neointimal lesion formation. Circ Res 100:1579–1588.

Kasim V, Miyagishi M, Taira K. (2004) Control of siRNA expression using the Cre-loxP recombination system. Nucleic Acids Res 32:e66.

Kasim V, Miyagishi M, Taira K. (2003) Control of siRNA expression utilizing Cre-loxP recombination system. Nucleic Acids Res Suppl 3:255–256.

Kramer A. (1996) The structure and function of proteins involved in mammalian pre-mRNA splicing. Annu Rev Biochem 65:367–409.

Kuehbacher A, Urbich C, Zeiher AM, Dimmeler S. (2007) Role of Dicer and Drosha for endothelial microRNA expression and angiogenesis. Circ Res 101:59–68.

Kunath T, Gish G, Lickert H, Jones N, Pawson T, Rossant J. (2003) Transgenic RNA interference in ES cell-derived embryos recapitulates a genetic null phenotype. Nat Biotechnol 21:559–561.

Lee DY, Deng Z, Wang CH, Yang BB. (2007) MicroRNA-378 promotes cell survival, tumor growth, and angiogenesis by targeting SuFu and Fus-1 expression. Proc Natl Acad Sci USA 104:20350–20355.

Li Y, Song YH, Li F, Yang T, Lu YW, Geng YJ. (2009) MicroRNA-221 regulates high glucose-induced endothelial dysfunction. Biochem Biophys Res Commun 381:81–83.

Litz J, Krystal GW. (2006) Imatinib inhibits c-Kitinduced hypoxia-inducible factor-1 _ activity and vascular endothelial growth factor expression in small cell lung cancer cells. Mol Cancer Ther 5:1415–1422.

Lu Y, Thomson JM, Wong HY, Hammond SM, Hogan BL. (2007) Transgenic over-expression of the microRNA miR-17-92 cluster promotes proliferation and inhibits differentiation of lung epithelial progenitor cells. Dev Biol 310:442–453.

Mendell JT. (2008) miRiad roles for the miR-17–92 cluster in development and disease. Cell 133:217–222.

Mertens-Talcott SU, Chintharlapalli S, Li X, Safe S. (2007) The oncogenic microRNA-27a targets genes that regulate specificity protein transcription factors and the G2-M checkpoint in MDA-MB-231 breast cancer cells. Cancer Res 67:11001–11011.

Mougin A, Gottschalk A, Fabrizio P, Luhrmann R, Branlant C. (2002) Direct probing of RNA structure and RNA–protein interactions in purified HeLa cell's and yeast spliceosomal U4/U6.U5 Tri-snRNP particles. J Mol Biol 317:631–649.

Peli J, Schmoll F, Laurincik J, Brem G, Schellander K. (1996) Comparison of aggregation and injection techniques in producing chimeras with embryonic stem cells in mice. Theriogenology 45:833–842.

Peng S, York JP, Zhang P. (2006) A transgenic approach for RNA interference-based genetic screening in mice. Proc Natl Acad Sci USA 103:2252–2256.

Poliseno L, Tuccoli A, Mariani L, Evangelista M, Citti L, Woods K, Mercatanti A, Hammond S, Rainaldi G. (2006) MicroRNAs modulate the angiogenic properties of HUVECs. Blood 108:3068–3071.

Rhoads K, Arderiu G, Charboneau A, Hansen SL, Hoffman W, Boudreau N. (2005) A role for Hox A5 in regulating angiogenesis and vascular patterning. Lymphat Res Biol 3:240–252.

Roboz GJ, Giles FJ, List AF, Cortes JE, Carlin R, Kowalski M, Bilic S, Masson E, Rosamilia M, Schuster MW, Laurent D, Feldman EJ. (2006) Phase 1 study of PTK787/ZK 222584, a small molecule tyrosine kinase receptor inhibitor, for the treatment of acute myeloid leukemia and myelodysplastic syndrome. Leukemia 20:952–957.

Rubinson DA, Dillon CP, Kwiatkowski AV, Sievers C, Yang L, Kopinja J, Rooney DL, Ihrig MM, McManus MT, Gertler FB, Scott ML, Van Parijs L. (2003) A lentivirus-based system to functionally silence genes in primary mammalian cells, stem cells and transgenic mice by RNA interference. Nat Genet 33:401–406.

Samols MA, Skalsky RL, Maldonado AM, Riva A, Lopez MC, Baker HV, Renne R. (2007) Identification of cellular genes targeted by KSHVencoded microRNAs. PLoS Pathog 3:e65.

Shilo S, Roy S, Khanna S, Sen CK. (2008) Evidence for the involvement of miRNA in redox regulated angiogenic response of human microvascular endothelial cells. Arterioscler Thromb Vasc Biol 28:471–477.

Sonnichsen B, Koski LB, Walsh A, Marschall P, Neumann B, Brehm M, Alleaume AM, Artelt J, Bettencourt P, Cassin E, Hewitson M, Holz C, Khan M, Lazik S, Martin C, Nitzsche B, Ruer M, Stamford J, Winzi M, Heinkel R, Roder M, Finell J, Hantsch H, Jones SJ, Jones M, Piano F, Gunsalus KC, Oegema K, Gonczy P, Coulson A, Hyman AA, Echeverri CJ. (2005) Full-genome RNAi profiling of early embryogenesis in Caenorhabditis elegans. Nature 434:462–469.

Strumberg D (2005) Preclinical and clinical development of the oral multikinase inhibitor sorafenib in cancer treatment. Drugs Today (Barc) 41:773–784.

Suárez Y, Fernández-Hernando C, Yu J, Gerber SA, Harrison KD, Pober JS, Iruela-Arispe ML, Merkenschlager M, Sessa WC. (2008) Dicer-dependent endothelial microRNAs are necessary for postnatal angiogenesis. Proc Natl Acad Sci USA 105:14082–14087.

Suárez Y, Fernández-Hernando C, Pober JS, Sessa WC. (2007) Dicer dependent microRNAs regulate gene expression and functions in human endothelial cells. Circ Res 100:1164–1173.

Tiscornia G, Tergaonkar V, Galimi F, Verma IM. (2004) CRE recombinase-inducible RNA interference mediated by lentiviral vectors. Proc Natl Acad Sci USA 101:7347–7351.

Urbich C, Kuehbacher A, Dimmeler S. (2008) Role of microRNAs in vascular diseases, inflammation, and angiogenesis. Cardiovasc Res 79:581–588.

van Solingen C, Seghers L, Bijkerk R, Duijs JM, Roeten MK, van Oeveren-Rietdijk AM, Baelde HJ, Monge M, Vos JB, de Boer HC, Quax PH, Rabelink TJ, van Zonneveld AJ. (2009) Antagomir-mediated silencing of endothelial cell specific microRNA-126 impairs ischemia-induced angiogenesis. J Cell Mol Med 13(8A):1577-1585.

Ventura A, Meissner A, Dillon CP, McManus M, Sharp PA, Van Parijs L, Jaenisch R, Jacks T. (2004) Cre-lox-regulated conditional RNA interference from transgenes. Proc Natl Acad Sci USA 101:10380–10385.

Volinia S, Calin GA, Liu CG, Ambs S, Cimmino A, Petrocca F, Visone R, Iorio M, Roldo C, Ferracin M, Prueitt RL, Yanaihara N, Lanza G, Scarpa A, Vecchione A, Negrini M, Harris CC, Croce CM. (2006) A microRNA expression signature of human solid tumors defines cancer gene targets. Proc Natl Acad Sci USA 103:2257–2261.

Wang S, Aurora AB, Johnson BA, Qi X, McAnally J, Hill JA, Richardson JA, Bassel-Duby R, Olson EN. (2008) The endothelial-specific microRNA miR-126 governs vascular integrity and angiogenesis. Dev Cell 15:261–271.

Xiao J, Yang B, Lin H, Lu Y, Luo X, Wang Z. (2007) Novel approaches for gene-specific interference via manipulating actions of microRNAs: examination on the pacemaker channel genes HCN2 and HCN4. J Cell Physiol 212:285–292.

Xin M, Small EM, Sutherland LB, Qi X, McAnally J, Plato CF, Richardson JA, Bassel-Duby R, Olson EN. (2009) MicroRNAs miR-143 and miR-145 modulate cytoskeletal dynamics and responsiveness of smooth muscle cells to injury. Genes Dev 23:2166–2178.

Yang WJ, Yang DD, Na S, Sandusky GE, Zhang Q, Zhao G. (2005) Dicer is required for embryonic angiogenesis during mouse development. J Biol Chem 280:9330–9335.

Ying SY, Chang DC, Lin SL. (2008) The microRNA (miRNA): overview of the RNA genes that modulate gene function. Mol Biotechnol 38:257–268.

Ying SY, Lin SL (2004) Intron-derived microRNAs--fine tuning of gene functions. Gene 342:25–28.

Ying SY, Lin SL. (2006) Current perspectives in intronic micro RNAs (miRNAs). J Biomed Sci 13:5–15.

Zhang Y, Lu Y, Wang N, Lin H, Pan Z, Gao X, Zhang F, Zhang Y, Xiao J, Shan H, Luo X, Chen G, Qiao G, Wang Z, Yang B. (2008) Control of experimental atrial fibrillation by microRNA-328. Circulation (accepted)

CHAPTER 8

miRNAs in Neointimal Lesion Formation

Abstract: Several processes including endothelial angiogenesis, vascular neointimal lesion formation, vascular inflammation process, lipoprotein metabolism, and hypertension are critically involved in atherosclerosis. This chapter aims to introduce the role of miRNAs in neointimal formation. Neointimal formation is a common pathological lesion in diverse cardiovascular diseases occurring at sites of subclinical atherosclerosis but are also classical hallmarks of restenosis after stenting, angioplasty, endarterectomy, and arterial transplantation. Neointimal growth is the balance between proliferation and apoptosis of vascular smooth muscle cells. A number of miRNAs, miR-21, miR-143, miR-145, miR-221, and miR-222, have been demonstrated to play important role in neointimal formation. Their corresponding target genes have also been established.

INTRODUCTION

Neointimal formation is a common pathological lesion in diverse cardiovascular diseases and the pathological basis of proliferative vascular diseases sharing similar cellular events and molecular mechanisms with cancers. Vascular hyperplasia and neointimal lesion formation results from rapid proliferation and growth of vascular cells, generally occurring after nonspecific vascular injury. Neointimal lesions occur at sites of subclinical atherosclerosis but are also classical hallmarks of restenosis after stenting, angioplasty, endarterectomy, and arterial transplantation [Rivard & Andrés V, 2000; Muto *et al.*, 2007]. Neointimal growth is the balance between proliferation and apoptosis of vascular smooth muscle cells (VSMCs). The increased VSMC proliferation or the relative decreased VSMC apoptosis are responsible for neointimal lesion formation. VSMC phenotypic switching from a contractile state to a proliferative state has arterial-wide ramifications to neointimal lesions and atherosclerotic plaques. miRNAs may be a new therapeutic target for proliferative vascular diseases such as atherosclerosis, postangioplasty restenosis, transplantation arteriopathy, and stroke.

Ji *et al* [2007] reported that multiple miRNAs are aberrantly expressed in the vascular wall after angioplasty with the time course changes matching the complex process of neointimal lesion formation, in which multiple genes are accordingly deregulated, as assessed with miRNA microarray. Seven days after balloon injury, 113 of the 140 artery miRNAs were differentially expressed (60 miRNAs were upregulated, and 53 miRNAs were downregulated). At 14 days after injury, 110 of the 140 artery miRNAs were differentially expressed (63 up and 47 down), whereas 102 of the 140 artery miRNAs were differentially expressed (55 up and 47 down) at 28 days after angioplasty. Specifically, miR-125a, miR-125b, miR-133, miR-143, miR-145, miR-365 appear to be downregulated, and miR-21, miR-146, miR-214, and miR-352 were observed to be upregulated in neointimal formation models [Ji *et al.*, 2007]. This study is the first to indicate that multiple miRNAs are involved in neointimal lesion formation.

ROLE OF miR-21

In the study reported by Ji *et al* [2007], miR-21 was found to be one of the most upregulated miRNAs in the vascular wall of balloon-injured rat carotid arteries. In this model, knockdown of miR-21 using antisense oligonucleotide inhibited neointima formation after angioplasty *in vivo*. In cultured VSMCs, depletion of miR-21 decreased proliferation and increased apoptosis. Their data further identified PTEN (phosphatase and TENsin homolog) and Bcl-2 as target genes for miR-21 in VSMCs.

Subsequently, the same group demonstrated that miRNAs are aberrantly expressed in vascular smooth muscle cells (VSMCs) after treatment with hydrogen peroxide (H_2O_2). H_2O_2 up-regulated miR-21 expression [Lin *et al.*, 2009]. H_2O_2-induced VSMC apoptosis and death were increased by miR-21 antisense inhibitor and decreased by pre-miR-21. Programmed cell death 4 (PDCD4) was established to be a direct target of miR-21. miR-21-mediated protective effect on VSMC apoptosis and death was blocked via adenovirus-mediated overexpression of PDCD4 without the miR-21 binding site. Moreover, activator protein 1 (AP1) was a downstream signaling molecule of PDCD4 in miR-21-modulated VSMCs. The results suggest that miRNAs in VSMCs are sensitive to H_2O_2 stimulation. miRN-21 participates in H_2O_2-mediated gene regulation and cellular injury response through PDCD4 and the activator protein 1 pathway. miRNAs might play a role in vascular diseases related to ROS [Lin *et al.*, 2009].

Zhiguo Wang

ROLE OF miR-145

Still a study from Zhang's group [Cheng *et al.*, 2009] observed that miR-145, the most abundant miRNA in normal vascular walls and in freshly isolated VSMCs [Ji *et al.*, 2007], is significantly downregulated in the vascular walls with neointimal lesion formation and in cultured dedifferentiated VSMCs. In cultured rat VSMCs *in vitro* and in balloon-injured rat carotid arteries *in vivo*, VSMC differentiation marker genes such as SM alpha-actin, calponin, and SM-MHC are upregulated by miR-145 overexpression but are downregulated by the miR-145 antisense. They further identified that miR-145-mediated phenotypic modulation of VSMCs is through its target gene KLF5 and its downstream signaling molecule, myocardin. Restoration of miR-145 in balloon-injured arteries via Ad-miR-145 inhibits neointimal growth. They believe that miR-145 is a novel VSMC phenotypic marker and modulator that is able of controlling vascular neointimal lesion formation [Zhang, 2009].

ROLE OF miR-221 AND miR-222

Again from Zhang's laboratory [Liu *et al.*, 2009], the authors reported the time course and cellular distribution of miR-221 and miR-222 expression in rat carotid arteries after angioplasty. Their expression was found upregulated and localized in VSMCs in the injured vascular walls. In cultured VSMCs, miR-221 and miR-222 expression was increased by growth stimulators. Knockdown of miR-221 and miR-222 resulted in decreased VSMC proliferation *in vitro*. Using both gain-of-function and loss-of-function approaches, they found that p27(Kip1) and p57(Kip2) were 2 target genes that were involved in miR-221- and miR-222-mediated effect on VSMC growth. Knockdown of miR-221 and miR-222 in rat carotid arteries suppressed VSMC proliferation *in vivo* and neointimal lesion formation after angioplasty.

ROLE OF miR-143 AND miR-145

The role of miRNAs in cardiovascular diseases has been recently further confirmed by studying their expression in the vascular smooth muscle cell (VSMC) compartment. For example, the miR-143/miR-145 cluster has been demonstrated to be specifically expressed in SMCs [Elia *et al.*, 2009; Cordes *et al.*, 2009; Cheng *et al.*, 2009; Xin *et al.*, 2009; Boettger *et al.*, 2009]; their expression is controlled by SRF and is decreased during acute (restenosis) or chronic (atherosclerosis) stress. Knockout of miR-143/miR-145 induces defects in VSMC terminal differentiation which is reflected by a decreased capacity for vasoconstriction after vasopressor challenge [Elia *et al.*, 2009; Xin *et al.*, 2009; Boettger *et al.*, 2009]. The cytoskeletal apparatus is particularly affected by the knockout of miR-143/miR-145.

Serum response factor (SRF) and its coactivator, myocardin, regulate a cardiovascular-specific miRNA cluster encoding miR-143 and miR-145. They play a central role in the control of smooth muscle phenotypes by regulating the expression of cytoskeletal genes. To assess the functions of these miRNAs *in vivo*, Xin *et al* [2009] systematically deleted them singly and in combination in mice. Mice lacking both miR-143 and miR-145 are viable and do not display overt abnormalities in smooth muscle differentiation, although they show a significant reduction in blood pressure due to reduced vascular tone. Remarkably, however, neointima formation in response to vascular injury is profoundly impeded in mice lacking these miRNAs, due to disarray of actin stress fibers and diminished migratory activity of SMCs. These abnormalities reflect the regulation of a cadre of modulators of SRF activity and actin dynamics by miR-143 and miR-145.

Elia *et al* [2009] showed a critical role of the miR-143/miR-145 cluster in VSMC differentiation and vascular pathogenesis, also through the generation of a mouse model of miR-143 and miR-145 knockout (KO). They determined that the expression of miR-143 and -145 is decreased in acute and chronic vascular stress (transverse aortic constriction and in aortas of the ApoE KO mouse). In human aortic aneurysms, the expression of miR-143 and miR-145 was significantly decreased compared with control aortas. In addition, overexpression of miR-143 and miR-145 decreased neointimal formation in a rat model of acute vascular injury. An in-depth analysis of the miR-143/miR-145 KO mouse model showed that this miR cluster is expressed mostly in the VSMC compartment, both during development and postnatally, in vessels and VSMC-containing organs. Loss of miR-143 and miR-145 expression induces structural modifications of the aorta, because of an incomplete differentiation of VSMCs. Their results indicate that the miR-143/miR-145 gene cluster has a critical role during VSMC differentiation and strongly suggest its involvement in the reversion of the VSMC differentiation phenotype that occurs during vascular disease.

REFERENCES

Cheng Y, Liu X, Yang J, Lin Y, Xu DZ, Lu Q, Deitch EA, Huo Y, Delphin ES, Zhang C. (2009) MicroRNA-145, a novel smooth muscle cell phenotypic marker and modulator, controls vascular neointimal lesion formation. Circ Res 105:158–166.

Cordes KR, Sheehy NT, White MP, Berry EC, Morton SU, Muth AN, Lee TH, Miano JM, Ivey KN, Srivastava D. (2009) miR-145 and miR-143 regulate smooth muscle cell fate and plasticity. Nature 460:705–710.

Elia L, Quintavalle M, Zhang J, Contu R, Cossu L, Latronico MV, Peterson KL, Indolfi C, Catalucci D, Chen J, Courtneidge SA, Condorelli G. (2009) The knockout of miR-143 and -145 alters smooth muscle cell maintenance and vascular homeostasis in mice: correlates with human disease. Cell Death Differ 16:1590–1598.

Ji R, Cheng Y, Yue J, Yang J, Liu X, Chen H, Dean DB, Zhang C. (2007) MicroRNA expression signature and antisense-mediated depletion reveal an essential role of MicroRNA in vascular neointimal lesion formation. Circ Res 100:1579–1588.

Lin Y, Liu X, Cheng Y, Yang J, Huo Y, Zhang C. (2009) Involvement of MicroRNAs in hydrogen peroxidemediated gene regulation and cellular injury response in vascular smooth muscle cells. J Biol Chem 284:7903–7913.

Liu X, Cheng Y, Zhang S, Lin Y, Yang J, Zhang C. (2009) A necessary role of miR-221 and miR-222 in vascular smooth muscle cell proliferation and neointimal hyperplasia. Circ Res 104:476–487.

Muto A, Fitzgerald TN, Pimiento JM, Maloney SP, Teso D, Paszkowiak JJ, Westvik TS, Kudo FA, Nishibe T, Dardik A. (2007) Smooth muscle cell signal transduction: implications of vascular biology for vascular surgeons. J Vasc Surg 45:A15–A24.

Rivard A, Andrés V. (2000) Vascular smooth muscle cell proliferation in the pathogenesis of atherosclerotic cardiovascular diseases. Histol Histopathol 15:557–571.

Xin M, Small EM, Sutherland LB, Qi X, McAnally J, Plato CF, Richardson JA, Bassel-Duby R, Olson EN. (2009) MicroRNAs miR-143 and miR-145 modulate cytoskeletal dynamics and responsiveness of smooth muscle cells to injury. Genes Dev 23:2166–2178.

Zhang C. (2009) MicroRNA-145 in vascular smooth muscle cell biology: a new therapeutic target for vascular disease. Cell Cycle 8:3469–3473.

Zhang C. (2009) MicroRNA and vascular smooth muscle cell phenotype: new therapy for atherosclerosis? Genome Med 1:85.

CHAPTER 9

miRNAs in Inflammatory Macrophages

Abstract: Several processes including endothelial angiogenesis, vascular neointimal lesion formation, vascular inflammation process, lipoprotein metabolism, and hypertension are critically involved in atherosclerosis. This chapter aims to introduce the role of miRNAs in vascular inflammation process. Atherosclerosis is now widely accepted to be an inflammatory disease, characterized by degenerative as well as proliferative changes and extracellular accumulation of lipid and cholesterol, in which an ongoing inflammatory reaction plays an important role in both initiation and progression/destabilization, converting a chronic process into an acute disorder. In early atheromotous plaques, inflammatory macrophages strive to alleviate the subendothelial accumulation of modified lipoproteins carrying cholesterol esters. Consequently, the further recruitment and migration of cells induce chronic inflammation. To date, only a few miRNAs have been shown to be involved in the vascular inflammation process. These include miR-126 in regulating adhesion molecules, and miR-155 and miR-125a in regulating inflammatory cytokine.

INTRODUCTION

Atherosclerosis-induced subintimal thickening results from both cellular infiltration and acellular accumulation of basement membrane proteins. In early atheromotous plaques, inflammatory macrophages strive to alleviate the subendothelial accumulation of modified lipoproteins carrying cholesterol esters [Vickers & Remaley, 2010]. Consequently, the further recruitment and migration of cells induce chronic inflammation. Macrophages loaded with engulfed cholesterol have reduced mobility and ultimately the burden results in a complete gene profile and phenotypic conversion to foam cells [Rader & Puré, 2005]. Inflammation drives the formation, progression, and rupture of atherosclerotic plaques. Experimental studies have demonstrated that an inflammatory subset of monocytes/macrophages preferentially accumulate in atherosclerotic plaque and produce proinflammatory cytokines. T lymphocytes can contribute to inflammatory processes that promote thrombosis by stimulating production of collagen-degrading proteinases and the potent procoagulant tissue factor. Recent data link obesity, inflammation, and modifiers of atherosclerotic events, a nexus of growing clinical concern given the worldwide increase in the prevalence of obesity. Modulators of inflammation derived from visceral adipose tissue evoke production of acute phase reactants in the liver, implicated in thrombogenesis and clot stability. Additionally, C-reactive protein levels rise with increasing levels of visceral adipose tissue. Adipose tissue in obese mice contains increased numbers of macrophages and T lymphocytes, increased T lymphocyte activation, and increased interferon-gamma (IFN-gamma) expression. IFN-gamma deficiency in mice reduces production of inflammatory cytokines and inflammatory cell accumulation in adipose tissue. Another series of *in vitro* and *in vivo* mouse experiments affirmed that adiponectin, an adipocytokine, the plasma levels of which drop with obesity, acts as an endogenous antiinflammatory modulator of both innate and adaptive immunity in atherogenesis. Thus, accumulating experimental evidence supports a key role for inflammation as a link between risk factors for atherosclerosis and the biology that underlies the complications of this disease. Evidence suggests that the vascular inflammation process is also regulated by miRNAs through targeting adhesion molecules and inflammatory cytokines.

ROLE OF miR-126 AND ADHESION MOLECULE

Adhesion molecules expressed by activated endothelial cells play a key role in regulating leukocyte trafficking to sites of inflammation. Resting endothelial cells normally do not express adhesion molecules, but cytokines activate endothelial cells to express adhesion molecules such as vascular cell adhesion molecule 1 (VCAM-1), which mediate leukocyte adherence to endothelial cells. Harris *et al* [2008] demonstrated that VCAM-1, which mediates leucocyte adherence to endothelial cells, is regulated by miR-126. Transfection of endothelial cells with an antisense oligo (AMO) to knockdown miR-126 results in an increase in TNFα-stimulated VCAM-1 expression [Harris *et al.*, 2008]. Conversely, overexpression of pre-miR-126 to increase cellular miR-126 level decreases VCAM-1 expression. Additionally, decreasing endogenous miR-126 levels increases leukocyte adherence to endothelial cells. These data suggest that miRNAs can regulate adhesion molecule expression and may provide additional control of vascular inflammation.

ROLE OF miR-155 AND INFLAMMATORY CYTOKINE

Recently, miR-155 has been implicated in down-modulating the inflammatory cytokine production in response to lipopolysaccharides in human monocyte-derived dendritic cells [Tili *et al.*, 2007]. miR-155 most probably directly targets transcript coding for several proteins involved in LPS signaling such as the Fas-associated death domain protein (FADD), IkappaB kinase epsilon (IKKepsilon), and the receptor (TNFR superfamily)-interacting serine-threonine kinase 1 (Ripk1) while enhancing TNF-alpha translation. Similar inflammatory response with miR-155 was also reported by another study [O'Connell *et al.*, 2007].

Worm *et al* [2009] showed that exposure of cultured macrophages and mice to lipopolysaccharide (LPS) leads to up-regulation of miR-155 and that the transcription factor c/ebp Beta is a direct target of miR-155. Interestingly, expression profiling of LPS-stimulated macrophages combined with overexpression and silencing of miR-155 in murine macrophages and human monocytic cells uncovered marked changes in the expression of granulocyte colony-stimulating factor (G-CSF), a central regulator of granulopoiesis during inflammatory responses. Regulation of inflammatory cytokine production by miR-155 via targeting C/EBPbeta was found in tumor-associated macrophages [He *et al.*, 2009].

ROLE OF miR-125a AND INFLAMMATORY CYTOKINE

Significant differential expression of miRNAs was observed in human peripheral blood monocytes treated with oxidized-LDL [Chen *et al.*, 2009]; specifically, miR-125a-5p, miR-146a, miR-146b-5p, miR-155 and miR-9 were all significantly upregulated. Inhibition of endogenous miR-125a-5p levels in THP-1 cells significantly increased the secretion of inflammatory cytokines (TGFβ, TNFα, IL-2, and IL-6) and increased the expression of macrophage scavenger receptors (LOX-1 and CD68), which resulted in increased lipid uptake [Chen *et al.*, 2009]. These data suggest that miR-125a-5p is antiatherogenic in the macrophage.

REFERENCES

Chen T, Huang Z, Wang L, Wang Y, Wu F, Meng S, Wang C. (2009) MicroRNA-125a-5p partly regulates the inflammatory response, lipid uptake, and ORP9 expression in oxLDL-stimulated monocyte/macrophages. Cardiovasc Res 83:131–139.

Harris TA, Yamakuchi M, Ferlito M, Mendell JT, Lowenstein CJ. (2008) MicroRNA-126 regulates endothelial expression of vascular cell adhesion molecule 1. Proc Natl Acad Sci USA 105:1516–21.

He M, Xu Z, Ding T, Kuang DM, Zheng L. (2009) MicroRNA-155 regulates inflammatory cytokine production in tumor-associated macrophages via targeting C/EBPbeta. Cell Mol Immunol 6:343–352.

Ji R, Cheng Y, Yue J, Yang J, Liu X, Chen H, Dean DB, Zhang C. (2007) MicroRNA expression signature and antisense-mediated depletion reveal an essential role of MicroRNA in vascular neointimal lesion formation. Circ Res 100:1579–1588.

Li Y, Song YH, Li F, Yang T, Lu YW, Geng YJ. (2009) MicroRNA-221 regulates high glucose-induced endothelial dysfunction. Biochem Biophys Res Commun 381:81–83.

Martin MM, Buckenberger JA, Jiang J, Malana GE, Nuovo GJ, Chotani M, Feldman DS, Schmittgen TD, Elton TS. (2007) The human angiotensin II type 1 receptor +1166 A/C polymorphism attenuates microrna-155 binding. J Biol Chem;282: 24262–24269.

O'Connell RM, Taganov KD, Boldin MP, Cheng G, Baltimore D. (2007) MicroRNA-155 is induced during the macrophage inflammatory response. Proc Natl Acad Sci USA 104:1604–1609.

Poliseno L, Tuccoli A, Mariani L, Evangelista M, Citti L, Woods K, Mercatanti A, Hammond S, Rainaldi G. (2006) MicroRNAs modulate the angiogenic properties of HUVECs. Blood 108:3068–3071.

Rader DJ, Puré E. (2005) Lipoproteins, macrophage function, and atherosclerosis: beyond the foam cell? Cell Metab 1:223–230.

Tili E, Michaille JJ, Cimino A, Costinean S, Dumitru CD, Adair B, Fabbri M, Alder H, Liu CG, Calin GA, Croce CM. (2007) Modulation of miR-155 and miR-125b levels following lipopolysaccharide/TNF-alpha stimulation and their possible roles in regulating the response to endotoxin shock. J Immunol 179:5082–5089.

Urbich C, Kuehbacher A, Dimmeler S. (2008) Role of microRNAs in vascular diseases, inflammation, and angiogenesis. Cardiovasc Res 79:581–588.

Vickers KC, Remaley AT. (2010) MicroRNAs in atherosclerosis and lipoprotein metabolism. Curr Opin Endocrinol Diabetes Obes 17:150–155.

Worm J, Stenvang J, Petri A, Frederiksen KS, Obad S, Elmén J, Hedtjärn M, Straarup EM, Hansen JB, Kauppinen S. (2009) Silencing of microRNA-155 in mice during acute inflammatory response leads to derepression of c/ebp Beta and down-regulation of G-CSF. Nucleic Acids Res 37:5784–5792.

Xin M, Small EM, Sutherland LB, Qi X, McAnally J, Plato CF, Richardson JA, Bassel-Duby R, Olson EN. (2009) MicroRNAs miR-143 and miR-145 modulate cytoskeletal dynamics and responsiveness of smooth muscle cells to injury. Genes Dev 23:2166–2178.

CHAPTER 10

miRNAs in Lipoprotein Metabolism

Abstract: Several processes including endothelial angiogenesis, vascular neointimal lesion formation, vascular inflammation process, lipoprotein metabolism, and hypertension are critically involved in atherosclerosis. This chapter aims to introduce the role of miRNAs in lipoprotein metabolism. The low-density lipoprotein (LDL) when oxidized by oxygen free radicals and coming into contact with arterial wall, causes atherosclerotic lesions leading to increase in endothelial permeability and adhesiveness. In response to the damage to the artery wall and endothelial dysfunction, the immune system responds by recruiting white blood cells to adsorb the oxidized-LDL. The stimulation of lipid uptake into these cells by oxidized-LDL is critical to the initiation and development of atherosclerosis. miR-122 has been documented to regulate cholesterol synthesis and miR-125a to regulate lipid uptake in monocytes/macrophages.

INTRODUCTION

Atherogenesis, also called atherosclerosis is defined as a process of forming atheromas, plaques in the inner lining of arteries driven, in part, by chronic inflammation in response to cholesterol accumulation in the arterial wall [Ross, 1999]. The first major event in the progression of the early atheroma is the loss of endothelial integrity; endothelium dysfunction facilitates the subendothelial accumulation of cholesterol-bearing lipoproteins. The low-density lipoprotein (LDL) when oxidized by oxygen free radicals (reactive oxygen species) and coming into contact with arterial wall, causes atherosclerotic lesions leading to increase in endothelial permeability and adhesiveness [Urbich *et al.*, 2008]. In response to the damage to the artery wall and endothelial dysfunction, the immune system responds by recruiting white blood cells (macrophages and T-lymphocytes) to adsorb the oxidized-LDL. The complications of advanced atherosclerosis are chronic, slowly progressive and cumulative, and result in thrombosis that decreases or stops blood flow. The coronary thrombosis causing heart attack and thrombosis in the arteries to the brain causing stroke are deadly forms of the disease. All cells synthesize cholesterol; however, the epicenter of lipoprotein metabolism resides in the liver. The hepatic cholesterol biosynthetic pathway is unique in its role in overall homeostasis of systemic cholesterol.

In addition, the inflammatory responses of monocytes/macrophages and the stimulation of lipid uptake into these cells by oxidized-LDL are critical to the initiation and development of atherosclerosis. Macrophages can take up oxidized-LDL, which leads to their conversion into foam cells [Osterud & Bjorklid, 2003; Shashkin *et al.*, 2005; Tabas, 2005], and these foam cells can in turn secrete many proinflammatory factors, such as transforming growth factor (TNF)-α and interleukins (ILs) [Chen *et al.*, 2006]. Cholesterol loaded macrophage foam cells are the hallmark of early atherosclerosis and eventually undergo secondary necrosis to form the lipid core of advanced atherosclerotic plaques.

ROLE OF miR-122 IN CHOLESTEROL SYNTHESIS

Recent studies have established the role of miRNA studies as modulators of the hepatic cholesterol network by directly regulating cholesterol synthesis genes (mRNA) or indirectly through one of the other co-ordinated pathways [Girard *et al.*, 2008; Lynn, 2009; Chen, 2009b].

Hepatic miRNA profiles show that miR-122 is disproportionately more abundant than any other hepatic miRNA [Lagos-Quintana *et al.*, 2002]. Multiple studies have shown that miR-122 is intricately involved in lipid metabolism [Esau *et al.*, 2006; Krutzfeldt *et al.*, 2005; Elmén *et al.*, 2008]. Overexpression of miR-122, with adenovirus, has been shown to increase the abundance of cholesterol synthesis genes (Hmgrcs1, Sqle, Dhcr7) in the liver, and in turn drive cholesterol synthesis [Krutzfeldt *et al.*, 2005]. Multiple studies have investigated the loss of function or knockdown of miR-122 in the hepatocyte and/or liver; however, the direct targets of miR-122 that confers the assumed negative repression upon the cholesterol pathway are unknown. Consequences of miR-122 loss include a significant drop in cholesterol synthesis gene expression, the reduction of plasma and hepatic cholesterol content, as well as an observed decrease in fatty acid synthesis [Esau *et al.*, 2006]. Results from a separate study, which used antagomirs of miR-122 to knockdown endogenous levels in mice, showed that many targets of miR-122 were upregulated in response to miR-122 loss [Krutzfeldt *et al.*, 2005]. Cholesterol synthesis genes, which are not predicted targets of miR-122, were downregulated along with reduced serum cholesterol levels in antagomiR-122 mice [Krutzfeldt *et al.*, 2005].

Further evidence will be needed to resolve the exact mechanism whereby miR-122 regulates the cholesterol biosynthetic pathway. Future studies will likely characterize the role of other miRNAs in directly repressing cholesterol synthesis, as well as functional targeting of other hepatic cholesterol network components; such as lipoprotein receptors, scavenger receptors, and genes in the bile-acid pathway.

ROLE OF miR-125a IN LIPID UPTAKE IN MONOCYTES/MACROPHAGES

Chen *et al* [2009a] investigated miRNAs in monocytes/macrophages and their potential role in oxidized-LDL-stimulation of lipid uptake and other atherosclerotic responses. Microarrays were used to analyse the global expression of microRNAs in oxidized-LDL-stimulated human primary peripheral blood monocytes, followed by TaqMan real-time PCR verification. Five miRNAs (miR-125a-5p, miR-9, miR-146a, mR-146b-5p, and miR-155) were found aberrantly expressed after oxidized-LDL treatment of human primary monocytes. Bioinformatics analysis suggested that miR-125a-5p is related to a protein similar to ORP9 (oxysterol binding protein-like 9) and this was confirmed by a luciferase reporter assay. Further experiments revealed that miR-125a-5p mediated lipid uptake and decreased the secretion of some inflammatory cytokines (interleukin-2, interleukin-6, tumour necrosis factor-α, transforming growth factor-β) in oxidized-LDL-stimulated monocyte-derived macrophages. The authors concluded that miR-125a-5p regulates the proinflammatory response, lipid uptake, and expression of ORP9 in oxidized-LDL-stimulated monocyte/macrophages thereby the initiation and development of atherosclerosis [Chen *et al.*, 2009a].

REFERENCES

Berk BC, Fujiwara K, Lehoux S. (2007) ECM remodeling in hypertensive heart disease. J Clin Invest 117:568–575.

Chen JW, Chen YH, Lin SJ. (2006) Long-term exposure to oxidized low-density lipoprotein enhances tumor necrosis factor-alpha-stimulated endothelial adhesiveness of monocytes by activating superoxide generation and redox-sensitive pathways. Free Radic Biol Med 40:817–826.

Chen T, Huang Z, Wang L, Wang Y, Wu F, Meng S, Wang C.(2009a) MicroRNA-125a-5p partly regulates the inflammatory response, lipid uptake, and ORP9 expression in oxLDL-stimulated monocyte/macrophages. Cardiovasc Res 83:131–139.

Chen XM. (2009b) MicroRNA signatures in liver diseases. World J Gastroenterol 15:1665–1672.

Elmén J, Lindow M, Silahtaroglu A, Bak M, Christensen M, Lind-Thomsen A, Hedtjärn M, Hansen JB, Hansen HF, Straarup EM, McCullagh K, Kearney P, Kauppinen S. (2008) Antagonism of microRNA-122 in mice by systemically administered LNA-antimiR leads to up-regulation of a large set of predicted target mRNAs in the liver. Nucleic Acids Res 36:1153–1162.

Esau C, Davis S, Murray SF, Yu XX, Pandey SK, Pear M, Watts L, Booten SL, Graham M, McKay R, Subramaniam A, Propp S, Lollo BA, Freier S, Bennett CF, Bhanot S, Monia BP. (2006) miR-122 regulation of lipid metabolism revealed by *in vivo* antisense targeting. Cell Metab 3:87–98.

Girard M, Jacquemin E, Munnich A, Lyonnet S, Henrion-Caude A. (2008) miR-122, a paradigm for the role of microRNAs in the liver. J Hepatol 48:648–656.

Horn M, Remkes H, Strömer H, Dienesch C, Neubauer S. (2001) Chronic phosphocreatine depletion by the creatine analogue beta-guanidinopropionate is associated with increased mortality and loss of ATP in rats after myocardial infarction. Circulation 104:1844–1849.

Inoko M, Kihara Y, Morii I, Fujiwara H, Sasayama S. (1994) Transition from compensatory hypertrophy to dilated, failing left ventricles in Dahl salt-sensitive rats. Am J Physiol 267:H2471–H2482.

Krutzfeldt J, Rajewsky N, Braich R, Rajeev KG, Tuschl T, Manoharan M, Stoffel M. (2005) Silencing of microRNAs *in vivo* with 'antagomirs'. Nature 438:685–689.

Lagos-Quintana M, Rauhut R, Lendeckel W, Tuschl T. (2001) Identification of novel genes coding for small expressed RNAs. Science 294:853–858.

Lu H, Buchan RJ, Cook SA. (2010) MicroRNA-223 regulates Glut4 expression and cardiomyocyte glucose metabolism. Cardiovasc Res. 2010 Feb 17. [Epub ahead of print]

Lynn FC. (2009) Meta-regulation: microRNA regulation of glucose and lipid metabolism. Trends Endocrinol Metab 20:452–459.

Osterud B, Bjorklid E. (2003) Role of monocytes in atherogenesis. Physiol Rev 83:1069–1112.

Ross R. (1999) Atherosclerosis: an inflammatory disease. N Engl J Med 340:115–126.

Shashkin P, Dragulev B, Ley K. (2005) Macrophage differentiation to foam cells. Curr Pharm Des 11:3061–3072.

Stanley WC, Recchia FA, Lopaschuk GD. (2005) Myocardial substrate metabolism in the normal and failing heart. Physiol Rev 85:1093–1129.

Tabas I. (2005) Consequences and therapeutic implications of macrophage apoptosis in atherosclerosis: the importance of lesion stage and phagocytic efficiency. Arterioscler Thromb Vasc Biol 25:2255–2264.

Tian R, Abel ED. (2001) Responses of GLUT4-deficient hearts to ischemia underscore the importance of glycolysis. Circulation 103:2961–2966.

CHAPTER 11

miRNAs and Hypertension

Abstract: Several processes including endothelial angiogenesis, vascular neointimal lesion formation, vascular inflammation process, lipoprotein metabolism, and hypertension are critically involved in atherosclerosis. This chapter aims to introduce the role of miRNAs in hypertension. Angiotensin II, the major bioactive peptide of the renin–angiotensin system, plays a crucial role in controlling various cardiovascular diseases, especially hypertension. Endothelium-dependent nitric oxide (NO) formation has been reproducibly demonstrated to be reduced in patients with essential hypertension compared with normotensive control subjects. Formation of NO in endothelial cells depends on an adequate and continuing supply of its key substrate, L-arginine. Studies on miRNAs in hypertension have been rather sparse, though miRNAs have been reported to play a role in hypertension. In particular, miR-155 regulates angiotensin II type 1 receptor and miR-122 regulates L-arginine transport.

INTRODUCTION

Hypertension is a complex, multifactorial disorder with genetic, environmental, and demographic factors contributing to its prevalence and pathogenesis. The genetic element affecting blood pressure variation is estimated to range from 30 to 50%. Therefore, identifying gene variants that contribute to hypertension may not only provide better understanding of the pathophysiology of the disease but also elucidate the biochemical and physiological pathways that link various risk factors in hypertension.

miR-155 AND ANGIOTENSIN II TYPE 1 RECEPTOR

Angiotensin II, the major bioactive peptide of the renin–angiotensin system, plays a crucial role in controlling cardiovascular homeostasis and is strongly involved in various cardiovascular diseases, especially hypertension and heart failure. Most of the physiological and pathological effects of angiotensin II are mediated by the angiotensin II type 1 receptor (AT1R) [Higuchi *et al.*, 2007]. In a first study, Martin *et al* [2006] demonstrated that miR-155 specifically interacts with the 3'UTR of the human AT1 receptor (hAT1R) mRNA, thereby reducing the endogenous expression of the hAT1R and consequently angiotensin II signaling. The miR-155 target site harboured in the 3'UTR of hAT1R mRNA is highly conserved across species. However although the AT1R mRNA sequence complementary to the miR-155 seed sequence (7 base-pairing) is perfectly conserved in chimpanzees and dogs, it is not perfectly conserved in mouse and rat AT1R genes. This may implicate that miR-155-mediated translation repression of the AT1R could only occur in humans, possibly in chimpanzees and dogs, but not in rodents [Scalbert & Bril, 2008].

The hAT1R gene has been found to be highly polymorphic, and consequences of AT1R polymorphisms on the regulation by miR-155 of AT1R density were studied recently by the same authors [Martin *et al.*, 2006]. Single nucleotide polymorphisms (SNPs) that occur in the 3'UTR may affect gene regulation by interfering with post-transcriptional activity, such as protein binding, polyadenylation and especially miRNA binding. In this regard, it is interesting to note that a particular SNP (rs5186) has been described in which there is an A/C transversion at position 1166 in the 3'UTR of AT1R gene. Moreover, an increased frequency of the hAT1R +1166C-allele has been associated with cardiovascular diseases, possibly as a result of enhanced AT1R activity [Bonnardeaux *et al.*, 1994; Cameron *et al.*, 2006]. Computer alignment revealed that the +1166 A/C SNP occurs in the cis-response element where miR-155 was shown to interact with the hAT1R. Consequently, it has been shown that the hAT1R +1166C-allele results in augmented hAT1R expression because of the inability of miR-155 to inhibit translation of hAT1R mRNA as efficiently [Martin *et al.*, 2007]. In situ hybridization experiments demonstrated that mature miR-155 is abundantly expressed in the same vascular cell types as the AT1R, that is endothelial cells and VSMCs. Primary human VSMCs transfected with an antisense oligonucleotide to miR-155 display augmented hAT1R expression and enhanced angiotensin II-induced signaling [Martin *et al.*, 2006].

Interestingly, because miR-155 is located on chromosome 21, it has been hypothesized that the lower blood pressure observed in trisomy 21 could be partially explained by overexpression of miR-155 (triplicated in trisomy 21),

Zhiguo Wang

leading to underexpression of hAT1R. Indeed, Sethupathy *et al* [2007] have shown in fibroblasts from monozygotic twins discordant for trisomy 21 (i.e. one twin unaffected and the other with trisomy 21) and homozygotes for the 'normal' 1166A hAT1R allele, that miR155 is indeed significantly overexpressed while levels of AT1R protein are lower in trisomy 21 [Martin *et al.*, 2006].

miRNAs IN SPONTANEOUSLY HYPERTENSIVE RATS

Xu *et al* [2008] investigated expression of miRNAs in the heart of hypertension in spontaneously hypertensive rats (SHR). miR-1, miR-133a, miR-155 and miR-208 were selected as the candidate miRNAs potentially related to blood pressure. The expression levels of miR-1, miR-133a, miR-155 and miR-208 in the aorta of 4-, 8-, 16- and 24-week-old SHR and age-matched Wistar-Kyoto (WKY) rats were detected by real-time RT-PCR. The mRNA levels of angiotensin II receptor type 1 (AGTR1a), angiotensin II receptor associated protein (AGTRAP), divalent metal transporter 1 (DMT1), lowdensity lipoprotein-related protein 1B (LRP1B), fibroblast growth factor-7 (FGF-7), protocadherin 9 precursor (PCDH9), chloride channel protein 5 (CLCN-5), small conductance calcium activated potassium channel protein 3 (KCNN3) and thyroid hormone receptor associated protein 1 (THRAP1), which were predicted to be target genes of differentially expressed miRNAs, were further detected by real-time RT-PCR. miR-155 level was found significantly lower in aorta of 16-week-old SHR than that of age-matched WKY rats and it was negatively correlated to blood pressure. In both WKY rats and SHR, miR-208 was most abundantly expressed in 4-week-old rats, but declined significantly in 8-, 16- and 24-week-old. No difference in miR-208 levels was observed between age-matched SHR and WKY rats. Moreover, miR-208 expression in aorta was negatively correlated with blood pressure and age. Neither miR-1 nor miR-133a was differentially expressed in SHR and WKY rats in different age groups. But the mRNA levels of predicted target genes were not correlated to miR-155 or miR-208 levels [Xu *et al.*, 2008]. These results indicate that miR-155 is less expressed in the aorta of adult SHR compared with that of WKY rats and is negatively correlated with blood pressure, suggesting it is possibly involved in the development and pathologic progress of hypertension.

miR-122 AND L-ARGININE TRANSPORT

Endothelium-dependent nitric oxide (NO) formation has been reproducibly demonstrated to be reduced in patients with essential hypertension compared with normotensive control subjects. Formation of NO in endothelial cells depends on an adequate and continuing supply of its key substrate, L-arginine, and several cofactors [Yang & Kaye, 2006]. Similar to their hypertensive counterparts, normotensive individuals at high risk for the development of hypertension are characterized by impaired L-arginine transport, which may represent the link between a defective L-arginine/NO pathway and the onset of essential hypertension [Schlaich *et al.*, 2004]. Arginine transport by endothelial cells is predominantly mediated by the system carriers. At normal physiological concentrations of arginine, SLC7A1 is the predominant transport system in endothelial cells, accounting for 60~80% of the total carrier-mediated uptake activity [Closs *et al.*, 1997]. Yang *et al* [2007] identified a functionally relevant singlenucleotide polymorphism (SNP) at 3'UTR of SLC7A1, and showed that it might account for the apparent link between altered endothelial function, L-arginine and NO metabolism, and predisposition to essential hypertension. They later demonstrated that the major allele contains a consensus sequence for the transcription factor SP1 and binds to SP1; in contrast, the minor allele fails to bind to SP1 [Yang & Kaye, 2009]. Resequencing of the entire SLC7A1 coding sequence failed to find other informative polymorphisms, indicating that ss52051869 plays a key role in the biochemical and clinical association.

By computational analysis, these authors found that the short and long variants of the 3'UTR of SLC7A1 contain three and four potential miR-122 binding sites, respectively [Yang & Kaye, 2009]. Tthe minor allele is more frequently associated with SLC7A1 bearing a long 3'UTR, while the major allele is more likely to accompany a short 3'UTR only. As such, reporter genes containing the long 3'UTR from SLC7A1 showed much less gene expression than those containing short 3'UTR from SLC7A1, regardless of their allele status, suggesting that an alternative polyadenylation event and/or miRNA-122 binding sites may also play a role in controlling gene expression. It is therefore possible that binding of miR-122 to the 3'UTR may cause the depression of gene expression, contributing to the lesser level of SLC7A1 and the endothelial dysfunction seen in hypertensive subjects.

REFERENCES

Bonnardeaux A, Davies E, Jeunemaitre X, Fèry I, Charru A, Clauser E, Tiret L, Cambien F, Corvol P, Soubrier F. (1994) AngiotensinII type 1 receptor gene polymorphisms in human essential hypertension. Hypertension 24:63–69.

Cameron VA, Mocatta TJ, Pilbrow AP, Frampton CM, Troughton RW, Richards AM, Winterbourn CC. (2007) Angiotensin type-1 receptor A1166C gene polymorphism correlates with oxidative stress levels in human heart failure. Hypertension 2006, 47:1155–1161.

Closs EI, Graf P, Habermeier A, Cunningham JM, Forstermann U. (1997) Human cationic amino acid transporters hCAT-1, hCAT-2A, and hCAT-2B: three related carriers with distinct transport properties. Biochemistry 36:6462–6468.

Higuchi S, Ohtsu H, Suzuki H, Shirai H, Frank GD, Eguchi S. (2007) Angiotensin II signal transduction through the AT1 receptor: novel insights into mechanisms and pathophysiology. Clin Sci 2007, 112:417–428.

Martin MM, Lee EJ, Buckenberger JA, Schmittgen TD, Elton TS. (2006) MicroRNA-155 regulates human angiotensin II type 1 receptor expression in fibroblasts. J Biol Chem 281:18277–18284.

Martin MM, Buckenberger JA, Jiang J, Malana GE, Nuovo GJ, Chotani M, Feldman DS, Schmittgen TD, Elton TS. (2007) The human angiotensin II type 1 receptor +1166 a/C polymorphism attenuates microRNA-155 binding. J Biol Chem 282:24262–24269.

Scalbert E, Bril A. (2008) Implication of microRNAs in the cardiovascular system. Curr Opin Pharmacol 8:181–188.

Schlaich MP, Parnell MM, Ahlers BA, Finch S, Marshall T, Zhang WZ, Kaye DM. (2004) Impaired L-arginine transport and endothelial function in hypertensive and genetically predisposed normotensive subjects. Circulation 110:3680–3686.

Sethupathy P, Borel C, Gagnebin M, Grant GR, Deutsch S, Elton TS, Hatzigeorgiou AG, Antonarakis SE. (2007) Human microRNA-155 on chromosome 21 differentially interacts with its polymorphic target in the AGTR1 30 untranslated region: a mechanism for functional single-nucleotide polymorphisms related to phenotypes. Am J Hum Genet 81:405–413.

Xu CC, Han WQ, Xiao B, Li NN, Zhu DL, Gao PJ. (2008) Differential expression of microRNAs in the aorta of spontaneously hypertensive rats. Sheng Li Xue Bao 60:553–560.

Yang Z, Kaye DM. (2009) Mechanistic insights into the link between a polymorphism of the 3'UTR of the SLC7A1 gene and hypertension. Hum Mutat 30:328–333.

Yang Z, Venardos K, Jones E, Morris BJ, Chin-Dusting J, Kaye DM. (2007) Identification of a novel polymorphism in the 3'UTR of the L-arginine transporter gene SLC7A1: contribution to hypertension and endothelial dysfunction. Circulation 115:1269–1274.

miRNAs in Cardiac Fibrosis

Abstract: The goal of this chapter is to discuss the regulation of cardiac fibrosis by miRNAs. Cardiac myocytes are normally surrounded by a fine network of collagen fibers. In the normal heart, two thirds of the cell population is composed of nonmuscle cells, the majority of which are fibroblasts. Cardiac fibrosis is the result of both an increase in fibroblast proliferation and extracellular matrix (ECM) deposition. A growing body of evidence indicates that, along with cardiomyocytes hypertrophy, diffusion of interstitial fibrosis is a key pathologic feature of myocardial remodelling in a number of cardiac diseases of different (e.g. ischemic, hypertensive, valvular, genetic, and metabolic) origin. The extracellular matrix (ECM) is a dynamic microenvironment; changes within ECM constitute the second important myocardial adaptation that occurs during cardiac remodelling. A subset of miRNAs is enriched in cardiac fibroblasts compared to cardiomyocytes. A number of studies have demonstrated the involvement of miRNAs in regulating myocardial fibrosis in the settings of myocardial ischemia or mechanical overload. Some miRNAs (miR-208 and miR-21) have been shown to favor fibrogenesis, being profibrotic miRNAs. Others including miR-29, miR-133, miR-30c, and miR-590 have been demonstrated to produce inhibitory effects on fibrogenesis, being anti-fibrotic miRNAs.

INTRODUCTION

In tissues composed of post-mitotic cells, like heart, new cells cannot be regenerated; instead, fibroblasts proliferate to fill the gaps created due to removal of dead cells. In the normal heart, two thirds of the cell population is composed of nonmuscle cells, the majority of which are fibroblasts [Maisch, 1995; Manabe *et al.*, 2002]. Cardiac fibroblasts, along with cardiomyocytes, play an essential role in the progression of cardiac remodeling. Damaging insults evoke multiple signaling pathways that lead to coordinate and sequential gene regulation; the initial events lead to the activation of cardiac fibroblasts. Cardiac fibrosis is the result of both an increase in fibroblast proliferation and extracellular matrix (ECM) deposition. Cardiac myocytes are normally surrounded by a fine network of collagen fibers. Myocardial fibrosis is an established morphological feature of the structural myocardial remodeling that is a characteristic of all forms of cardiac pathology [Berk *et al.*, 2007; Khan & Sheppard, 2006]. A growing body of evidence indicates that, along with cardiomyocytes hypertrophy, diffuse interstitial fibrosis is a key pathologic feature of myocardial remodeling in a number of cardiac diseases of different (e.g. ischemic, hypertensive, valvular, genetic, and metabolic) origin. Acute myocardial infarction (MI) due to coronary artery occlusion represents a major cause of morbidity and mortality in humans [Fox *et al.*, 2007]. The loss of blood flow to the left ventricular free wall of the heart after MI results in death of cardiomyocytes and impaired cardiac contractility. Scar formation at the site of the infarct and interstitial fibrosis of adjacent myocardium prevent myocardial repair, diminish coronary reserve and contribute to loss of pump function, and predisposes individuals to ventricular dysfunction and arrhythmias, which, in turn, confer an increased risk of adverse cardiovascular events [Swynghedauw, 1999]. Fibrosis is also an important feature of cardiac adaptation to stress, which results in the abnormal myocardial stiffness that contributes to diastolic dysfunction, cardiomyocyte loss, arrhythmias, and the progression to heart failure.

The extracellular matrix (ECM) is a dynamic microenvironment; changes within ECM constitute the second important myocardial adaptation that occurs during cardiac remodeling. The myocardial ECM consists of a basement membrane, a fibrillar collagen network that surrounds myocytes, proteoglycans, and glycosaminoglycans, and biologically active signaling molecules. Important changes occur in the ECM during cardiac remodeling, including changes in fibrillar collagen synthesis and degradation, as well as changes in the degree of collagen cross-linking that contributes to adverse ventricular remodeling after MI and in advancing heart failure, via activation of profibrotic pathways and matrix metalloproteinases that enhance collagenase activity to increase in collagen content of the heart or myocardial fibrosis [Bouzeghrane *et al.*, 2005]. Excess fibrotic tissue results from altered regulation of the myocardial ECM, characterized by both increased synthesis and unchanged or decreased degradation of extracellular matrix proteins such as collagen types I and III. This imbalance is thought to be mediated by a combination of mechanical (e.g. cyclic stretch), neurohormonal (e.g. angiotensin II) and cytokine (e.g. transforming growth factor beta [TGF-β]) factors that act on cardiac cells. Phenotypically transformed fibroblast-like cells, termed myofibroblasts, are primarily responsible for fibrous tissue formation at the site of infarction [Powell *et al.*, 1999]

Zhiguo Wang

(Fig. **1**). In addition, myocardial fibrosis may provide the structural substrate for atrial and ventricular arrhythmias, thus potentially contributing to sudden death. Thus, reversal of this process represents an important therapeutic target in post-MI management and heart failure.

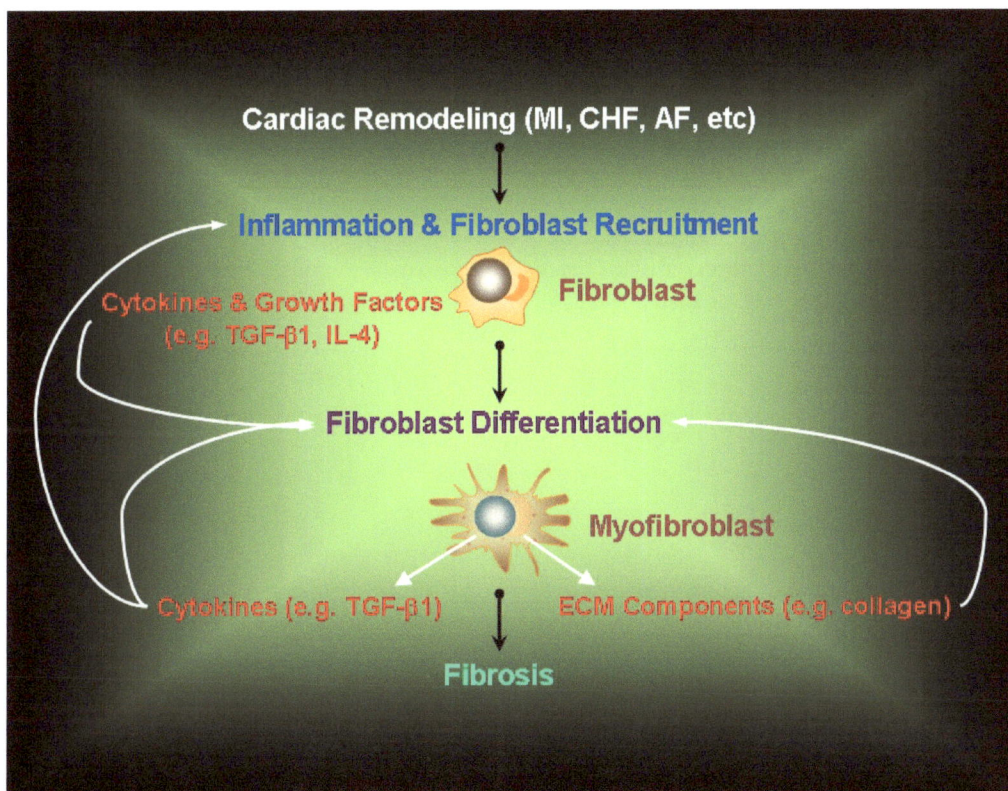

Figure 1: Diagram illustrating the pathways of cardiac fibrogenesis. Solid arrow indicates inductio. MI: myocardial ischemia; CHF: cardiac hypertrophy/heart failure; AF: atrial fibrillation.

Elucidation of the precise mechanisms responsible for the actions of these factors could forge new frontiers in both risk identification and prevention of fibrosis derived clinical complications in patients with cardiac disease. A subset of miRNAs is enriched in cardiac fibroblasts compared to cardiomyocytes, including miR-21 and members of the miR-29 family [Thum *et al*., 2008]. A number of studies have demonstrated the involvement of miRNAs in regulating myocardial fibrosis in the settings of myocardial ischemia or mechanical overload. In this conceptual framework, the investigation of miRNAs might offer a new opportunity to advance our knowledge of the pathogenesis of fibrosis. da Costa Martins *et al* [2008] reported that conditional deletion of Dicer in the adult mouse myocardium resulted in hypertrophic growth of cardiomyocytes, myocardial fibrosis, and functional defects in the ventricles. Characterization of individual miRNAs or miRNA expression profiles that are specifically associated with myocardial fibrosis might allow us to develop diagnostic tools and innovative therapies for fibrogenic cardiac diseases. The identification of miRNAs as potential regulators of myocardial fibrosis has clinical implications; the search for a miRNA expression pattern specific to fibrosis might provide a novel diagnostic approach. Yet, the molecular mechanisms that lead to a fibrogenic cardiac phenotype are still being elucidated.

PROFIBROTIC miRNAs

miR-208 in Hypertrophied/Failing Heart

Olson and colleagues have shown that miR-208, a cardiac-specific miRNA encoded within an intron of the αMHC gene, is required for cardiac fibrosis in response to hemodynamic pressure overload. Indeed, miR-208–deficient mice (miR-208$^{-/-}$) were resistant to developing fibrosis following TAB, as well as cardiac-restricted overexpression of calcineurin, a calcium/calmodulin-dependent phosphatase that provokes pathological remodeling of the heart [van Rooij *et al*., 2007]. Thus, miR-208 appears to be required for the development of cardiac fibrosis.

miR-21 in Ischemic Myocardium

Profiling of miRNA expression levels in the border zone of the infarct and remote myocardium both 3 days and 2 weeks post-MI revealed a striking miRNA expression pattern [van Rooij *et al*., 2008]. Among the regulated miRNAs, the miR-21 is dramatically upregulated in the border zone flanking the infracted area. Roy *et al* [2009] observed similar increase in miR-21 but in the infarct region of the IR heart. This result has also been reproduced by other studies [Cheng *et al*., 2007; Tatsuguchi *et al*., 2007]. Together with the finding that miR-21 functions as an oncogene and has a role in tumorigenesis by promoting cell proliferation [Tong & Nemunaitis, 2008] has led to the speculation that the induction of miR-21 in fibroblasts of diseased hearts might contribute, at least partially, to the increase in fibroblast proliferation. Indeed, studies with isolated cardiac fibroblasts demonstrated that phosphatase and tensin homolog (PTEN) is a direct target of miR-21 [Roy *et al*., 2009]. Modulation of miR-21 regulated expression of matrix metalloproteases 2 (MMP-2) via a PTEN pathway. Consistely, a marked decrease in PTEN expression in the infarct zone was observed. This decrease was associated with increased MMP2 expression in the infarct area (Fig. **2**). These data indicate that miR-21 is a profibrotic miRNA acting on ECM proteins.

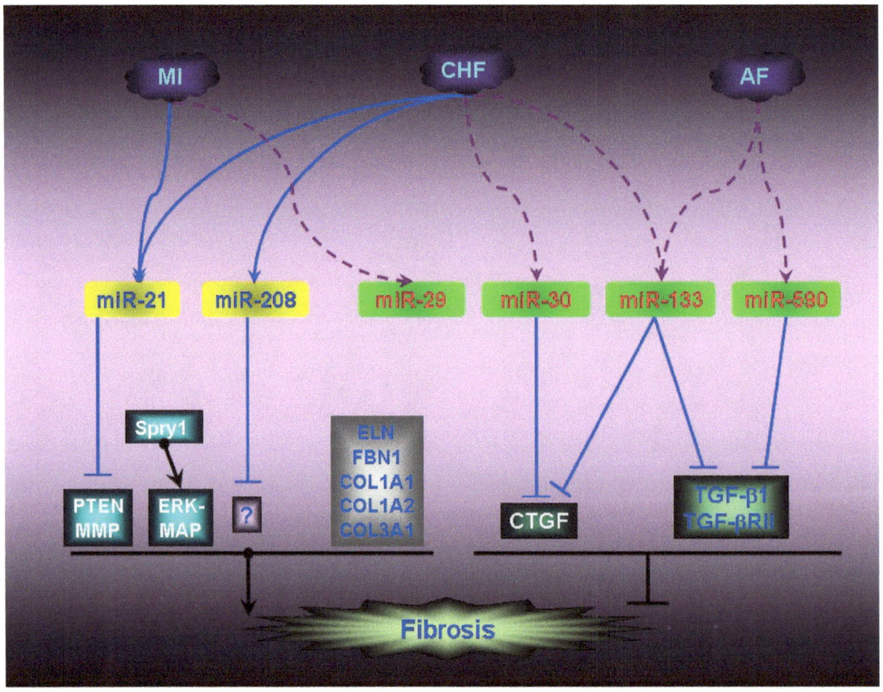

Figure 2: Diagram illustrating the role and signaling mechanisms of miRNAs in cardiac fibrosis. Solid arrow indicates induction; dashed line indicates inhibition; and pause sign indicates repression. MI: myocardial ischemia; CHF: cardiac hypertrophy/heart failure; AF: atrial fibrillation.

MMPs are best known for their actions in extracellular matrix (ECM) remodelling, but evidence is accumulating, especially for MMP-2 that is expressed ubiquitously in cardiomyocytes, endothelial cells and fibroblasts, that they may also play an important role intracellularly, particularly in response to I/R-I [Chow *et al*., 2007]. Myocardial stunning in response to I/R-I activates MMPs and inactivate their endogenous inhibitors, the tissue inhibitors of metalloproteinases (TIMPs). The imbalance between TIMPs and MMPs in the heart may be one of the contributing factors to acute I/R-I [Chow *et al*., 2007].

miR-21 has also been reported to be progressively and selectively upregulated in cardiac fibroblasts during the later stages of heart failure [Thum *et al*., 2008]. This increase results in augmentation of ERK-MAP kinase activity through inhibition of sprouty homologue 1 (Spry1). This mechanism regulates fibroblast survival and growth factor secretion, apparently controlling the extent of interstitial fibrosis and cardiac hypertrophy. *In vivo* silencing of miR-21 by a specific antagomir in a mouse pressure-overload-induced disease model reduces cardiac ERK-MAP kinase activity, inhibits interstitial fibrosis and attenuates cardiac dysfunction. These findings reveal that miRNAs can contribute to myocardial disease by an effect in cardiac fibroblasts.

ANTIFIBROTIC miRNAs

miR-29 in Ischemic Myocardium

van Rooij *et al* [2008] provided the first documentation about the role of miRNA in myocardial fibrosis by revealing miR-29 as a anit-fibrotic miRNA in myocardium. The miR-29 family of miRNAs (miR-29a, miR-29b, miR-29c) are preferentially expressed in the fibroblast population of the myocardium and was found downregulated in areas surrounding infarcted areas in mouse and human hearts, along with downregulation of miR-499 and upregulation of miR-21, miR-214, and miR-223 [van Rooij *et al.*, 2008]. The miR-29 family targets mRNAs that encode various extracellular matrix proteins involved in fibrosis, including elastin (ELN), fibronectin-1 (FBN1), collagen-1A1 (COL1A1), COL1A2 and COL3A1. They also showed that TGF-β decreases miR-29 expression in rat fibroblasts [van Rooij *et al.*, 2008]. Artificial overexpression of miR-29b in fibroblasts, achieved by use of a miR-29 mimic, resulted in reduced collagen expression. Conversely, downregulation of miR-29b by its use of anti-miR both *in vitro* and *in vivo* induced expression of collagens. Hence, reduction of miR-29 expression or its activity may contribute to fibrosis in the ischemic myocardium. Importantly, miR-29 is also downregulated in disease models for cardiac hypertrophy and failure [van Rooij *et al.*, 2006]. It is therefore quite plausible that downregulation of miR-29 also accounts at least partially for the abnormal fibrogenesisin these pathological settings. These data imply that therapeutic upregulation of miR-29 in response to an ischemic event or cardiac stress might prevent the onset of cardiac fibrosis and thereby maintain cardiac function. But as noticed by the authors, although their results suggest a key role for miR-29 in the control of cardiac fibrosis, there is not a direct one-to-one stoichiometric relationship between the levels of miR-29 and collagen expression. A fewfold decrease in miR-29 expression after MI was accompanied by up to a 20-fold increase in collagen mRNA expression, whereas a miR-29 antagomir induced collagens only slightly.

Role of miR-133 and miR-30c in Failing Heart

Connective tissue growth factor (CTGF) is a key molecule in the process of fibrosis and an attractive therapeutic target for cardiac disease associated with abnormal fibrosis. Regulation of CTGF expression at the transcription level has been studied extensively. A recent study showed that CTGF is also importantly regulated at the post-transcriptin level by 2 major cardiac miRNAs, miR-133 and miR-30 [Duisters *et al.*, 2009]. The expression of both miRNAs was found inversely related to the amount of CTGF in 2 rodent models of heart disease and in human pathological left ventricular hypertrophy. In cultured cardiomyocytes and fibroblasts, knockdown of these miRNAs increased CTGF levels. Further, overexpression of miR-133 or miR-30c decreased CTGF levels, which was accompanied by decreased production of collagens. Moreover, CTGF is a direct target of these miRNAs, because they directly interact with the 3'untranslated region of CTGF. These results indicate that miR-133 and miR-30 importantly limit the production of CTGF. Finally, the authors also provided evidence that the decrease of miR-133 and miR-30 in pathological left ventricular hypertrophy allows CTGF levels to increase, which contributes to collagen synthesis.

In another study, Matkovich *et al* [2010] reported that preventing "normal" downregulation of miR-133a in actively hypertrophying hearts after pressure overloading decreases fibrotic myocardial remodeling and improve diastolic performance, without altering the extent of hypertrophy. The typical increase in left ventricular end-diastolic pressure after TAC was diminished with miR-133a overexpression, and pressure–volume analysis revealed improved myocardial stiffness in miR-133a hearts. This finding further confirmed that miR-133 is an antifibrotic miRNA.

Also consistent with the postulated role of miR-133 as a negative regulator of fibrosis, Olson's group reported that knockout mice lacking miR-133-a1 and miR-133-a2 develop severe fibrosis and heart failure, resulting in cardiomyopathy and sudden death [Liu *et al.*, 2008]. These phenotypes were ascribed to the upregulation of serum response factor, a known transcriptional activator of heart failure.

miR-133 and miR-590 in Atrial Fibrillation

A large number of people smoke cigarettes and/or use over-the-counter nicotine products (patches and gums) to satisfy nicotine addiction. Serious, sometimes fatal, cases of atrial fibrillation (AF) have been reported in patients with and without detectable structural and/or history of heart disease from using a nicotine product, particularly when patients have smoked while using nicotine patches. In a study by Shan *et al* [2009] to elucidate molecular mechanisms underlying nicotine's promoting AF by inducing atrial structural remodeling, they found that nicotine stimulated remarkable collagen production and atrial fibrosis both *in vitro* in cultured canine atrial fibroblasts and *in vivo* in canine atrium subjected to rapid atrial pacing to induce AF. Nicotine produced significant upregulation of expression of TGF-β1 and TGFβRII at the protein level, and a 60–70% decrease in the levels of miRNAs miR-133 and miR-590. This downregulation of miR-133 and miR-590 partly accounts for the upregulation of TGF-β1 and TGFβRII, because our data established TGF-β1 and TGF-βRII as targets for miR-133 and miR-590 repression. Transfection of miR-133 or miR-590 into cultured atrial fibroblasts decreased TGF-β1 and TGF-βRII levels and collagen content. These effects were abolished by the antisense oligonucleotides against miR-133 or miR-590. The authors concluded that the profibrotic response to nicotine in canine atrium is critically dependent upon downregulation of antifibrotic miRNAs miR-133 and miR-590.

REFERENCES

Babij P, Askew GR, Nieuwenhuijsen B, Su CM, Bridal TR, Jow B, Argentieri TM, Kulik J, DeGennaro LJ, Spinelli W, Colatsky TJ. (1998) Inhibition of cardiac delayed rectifier K$^+$ current by overexpression of the long-QT syndrome HERG G628S mutation in transgenic mice. Circ Res 83:668–678.

Berk BC, Fujiwara K, Lehoux S. (2007) ECM remodeling in hypertensive heart disease. J Clin Invest 117:568–575.

Bouzeghrane F, Reinhardt DP, Reudelhuber TL, Thibault G. (2005) Enhanced expression of fibrillin-1, a constituent of the myocardial extracellular matrix in fibrosis. Am J Physiol Heart Circ Physiol 289:H982–991.

Cheng Y, Ji R, Yue J, Yang J, Liu X, Chen H, Dean DB, Zhang C. (2007) MicroRNAs are aberrantly expressed in hypertrophic heart. Do they play a role in cardiac hypertrophy? Am J Pathol 170:1831–1840.

Chow AK, Cena J, Schulz R. (2007) Acute actions and novel targets of matrix metalloproteinases in the heart and vasculature. Br J Pharmacol 152:189–205.

Costinean S, Zanesi N, Pekarsky Y, Tili E, Volinia S, Heerema N, Croce CM. (2006) Pre-B cell proliferation and lymphoblastic leukemia/high-grade lymphoma in E(mu)-miR155 transgenic mice. Proc Natl Acad Sci USA 103:7024–7029.

da Costa Martins PA, Bourajjaj M, Gladka M, Kortland M, van Oort RJ, Pinto YM, Molkentin JD, De Windt LJ. (2008) Conditional dicer gene deletion in the postnatal myocardium provokes spontaneous cardiac remodeling. Circulation 118:1567–1576.

Duisters RF, Tijsen AJ, Schroen B, Leenders JJ, Lentink V, van der Made I, Herias V, van Leeuwen RE, Schellings MW, Barenbrug P, Maessen JG, Heymans S, Pinto YM, Creemers EE. (2009) mir-133 and mir-30 regulate connective tissue growth factor: implications for a role of microrNAs in myocardial matrix remodeling. Circ Res 104:170–178.

Everett CA, West JD. (1996) The influence of ploidy on the distribution of cells in chimaeric mouse blastocysts. Zygote 4:59–66.

Fox CS, Coady S, Sorlie PD, D'Agostino RB Sr, Pencina MJ, Vasan RS, Meigs JB, Levy D, Savage PJ. (2007) Increasing cardiovascular disease burden due to diabetes mellitus: The Framingham Heart Study. Circulation 115:1544–1550.

Khan R, Sheppard R. (2006) Fibrosis in heart disease: understanding the role of transforming growth factor-beta in cardiomyopathy, valvular disease and arrhythmia. Immunology 118:10–24.

Kunath T, Gish G, Lickert H, Jones N, Pawson T, Rossant J. (2003) Transgenic RNA interference in ES cell-derived embryos recapitulates a genetic null phenotype. Nat Biotechnol 21:559–561.

Liu N, Bezprozvannaya S, Williams AH, Qi X, Richardson JA, Bassel-Duby R, Olson EN. (2008) microRNA-133a regulates cardiomyocyte proliferation and suppresses smooth muscle gene expression in the heart. Genes Dev 22:3242–3254.

Lu Y, Thomson JM, Wong HY, Hammond SM, Hogan BL. (2007) Transgenic over-expression of the microRNA miR-17-92 cluster promotes proliferation and inhibits differentiation of lung epithelial progenitor cells. Dev Biol 310:442–453.

Maisch B. (1995) Extracellular matrix and cardiac interstitium: restriction is not a restricted phenomenon. Herz 20:75–80.

Manabe I, Shindo T, Nagai R. (2002) Gene expression in fibroblasts and fibrosis involvement in cardiac hypertrophy. Circ Res 91:1103–1113.

Martin MM, Lee EJ, Buckenberger JA, Schmittgen TD, Elton TS. (2006) MicroRNA-155 regulates human angiotensin II type 1 receptor expression in fibroblasts. J Biol Chem 281:18277–18284.

Matkovich SJ, Wang W, Tu Y, Eschenbacher WH, Dorn LE, Condorelli G, Diwan A, Nerbonne JM, Dorn GW 2nd. (2010) MicroRNA-133a protects against myocardial fibrosis and modulates electrical repolarization without affecting hypertrophy in pressure-overloaded adult hearts. Circ Res 106:166–175.

Peli J, Schmoll F, Laurincik J, Brem G, Schellander K. (1996) Comparison of aggregation and injection techniques in producing chimeras with embryonic stem cells in mice. Theriogenology 45:833–842.

Powell DW, Mifflin RC, Valentich JD, Crowe SE, Saada JI, West AB. (1999) Myofibroblasts. I. Paracrine cells important in health and disease. Am J Physiol 277:C1–C9.

Roy S, Khanna S, Hussain SR, Biswas S, Azad A, Rink C, Gnyawali S, Shilo S, Nuovo GJ, Sen CK. (2009) MicroRNA expression in response to murine myocardial infarction: miR-21 regulates fibroblast metalloprotease-2 via phosphatase and tensin homologue. Cardiovasc Res 82:21-29.

Shan H, Zhang Y, Lu Y, Zhang Y, Pan Z, Cai B, Wang N, Li X, Feng T, Hong Y, Yang B. (2009) Downregulation of miR-133 and miR-590 contributes to nicotine-induced atrial remodelling in canines. Cardiovasc Res 83:465–472.

Swynghedauw B. (1999) Molecular mechanisms of myocardial remodeling. Physiol Rev 79:215–262.

Tatsuguchi M, Seok HY, Callis TE, Thomson JM, Chen JF, Newman M, Rojas M, Hammond SM, Wang DZ. (2007) Expression of microRNAs is dynamically regulated during cardiomyocyte hypertrophy. J Mol Cell Cardiol 42:1137–1141.

Thum T, Catalucci D, Bauersachs J. (2008) MicroRNAs: novel regulators in cardiac development and disease. Cardiovasc Res 79:562–570.

Thum T, Gross C, Fiedler J, Fischer T, Kissler S, Bussen M, Galuppo P, Just S, Rottbauer W, Frantz S, Castoldi M, Soutschek J, Koteliansky V, Rosenwald A, Basson MA, Licht JD, Pena JT, Rouhanifard SH, Muckenthaler MU, Tuschl T, Martin GR, Bauersachs J, Engelhardt S. (2008) MicroRNA-21 contributes to myocardial disease by stimulating MAP kinase signalling in fibroblasts. Nature 456:980–984.

Tong AW, Nemunaitis J. (2008) Modulation of miRNA activity in human cancer: a new paradigm for cancer gene therapy? Cancer Gene Ther 15:341–355.

van Rooij E, Sutherland LB, Qi X, Richardson JA, Hill J, Olson EN. (2007) Control of stress-dependent cardiac growth and gene expression by a microRNA. Science 316:575–579.

van Rooij E, Sutherland LB, Thatcher JE, DiMaio JM, Naseem RH, Marshall WS, Hill JA, Olson EN. (2008) Dysregulation of microRNAs following myocardial infarction reveals a role of miR-29 in cardiac fibrosis. Proc Natl Acad Sci USA 105:13027–13032.

CHAPTER 13

miRNAs in Cardiomyocyte Apoptosis

Abstract: This chapter aims to discuss the regulation of cardiomyocyte apoptosis by miRNAs. Apoptosis is an active process that leads to cell death. Unlike necrosis, apoptosis is a complex endogenous gene-controlled event that requires an exogenous signal–stimulated or inhibited by a variety of regulatory factors, such as formation of oxygen free radicals, ischemia, hypoxia, reduced intracellular K^+ concentration, and generation of nitric oxide. Apoptosis has been implicated in a variety of human disease including heart disease, Alzheimer's disease, cancer, etc. To date, no less than 30 individual miRNAs are known to regulate apoptosis. The number in the list is expected to expand quickly with more studies. A number of miRNAs including miR-1, miR-29, and miR-320 are considered proapoptotic miRNAs. The miRNAs identified to date possessing antiapoptotic action include miR-133, miR-21 and miR-199a. In addition, miR-21 has also been shown to produce proproliferative and anti-apoptotic effects in vascular smooth muscle cells. This chapter provides detailed description of these individual miRNAs for their role in cardiomyocyte apoptosis.

INTRODUCTION

Apoptosis is an active process that leads to cell death. Unlike necrosis, apoptosis is a complex endogenous gene-controlled event that requires an exogenous signal–stimulated or inhibited by a variety of regulatory factors, such as formation of oxygen free radicals, ischemia, hypoxia, reduced intracellular K^+ concentration, and generation of nitric oxide. Apoptosis has been implicated in a variety of human disease including heart disease, Alzheimer's disease, cancer, etc. In addition to well-established features of apoptotic cells, such as cell shrinkage, membrane blebbing, chromatin condensation and DNA fragmentation, we have recently identified three intrinsic properties of apoptosis. The first of these is what we term Apoptotic Preconditioning: a brief exposure of cells to a sublethal dose of apoptotic inducers increases the tolerance of the cells to subsequent lethal apoptotic insults [Han *et al.*, 2001]. It is a powerful means of cytoprotection from various environmental and intracellular stresses. The second property is Apoptotic Remodeling: a brief exposure of cells to a lethal dose of apoptosis inducers can trigger a signaling transduction cascade which commits cells to death even though the inducers are withdrawn [Han *et al.*, 2004]. This property of apoptosis may contribute to anatomical remodeling occurring under several disease conditions, such as the transition from cardiac hypertrophy to heart failure and Alzheimer's disease, as well as to morphogenesis of the organism. Apoptotic Resistance is the third intrinsic property: cells surviving super-lethal apoptotic insults lose their sensitivity to apoptotic inducers [Lu *et al.*, 2008]. This property may contribute to the resistance of cells, such as tumor cells, to apoptotic inducers.

To date, no less than 30 individual miRNAs are known to regulate apoptosis. These include the let-7 family, miR-1, miR-1d, miR-7, miR-14, miR-15a, miR-16-1, miR-17 cluster (miR-17-5p, miR-18, miR-19a, miR-19b, miR-20 and miR-92), miR-21, miR-29, miR-34a, miR-133, miR-146a, miR-146b, miR-148, miR-191, miR-204, miR-210, miR-214, miR-216, miR-278, miR-296, miR-335, miR-Lat, and bantam. The number of the list is expected to expand quickly with more studies. Indeed, we have performed a bioinformatics prediction of the vertebrate miRNAs available to date in miRBase using a target scan computational analysis with miRBase and miRanda and surprisingly found that nearly all known vertebrate miRNAs (~93%) have at least one target gene related to cell death and survival or cell cycle and proliferation [Yang *et al.*, 2009]. This suggests that a majority of, if not all, vertebrate miRNAs can regulate apoptosis in at least some cell types.

Among the known apoptosis-regulating miRNAs, some are designated anti-apoptotic and others proapoptotic miRNAs. This distinction is primarily based upon experimental results from a particular cell type.

In general, the following miRNAs are considered anti-apoptotic: miR-17-5p, miR-20a, miR-21, miR-133, miR-146a, miR-146b, miR-191, miR-14, bantam, miR-1d, miR-7, miR-148, miR-204, miR-210, miR-216, miR-296, and miR-Lat. For example, inhibition of miR-17-5p and miR-20a with their antisense oligonucleotides (AMOs) induces apoptosis in lung cancer cells, indicating that miR-17-5p and miR-20a can protect these cells against apoptosis [Ji *et al.*, 2007]. Knockdown of miR-21 in cultured glioblastoma cells resulted in a significant drop in cell number. This reduction was accompanied by increases in caspase-3 and -7 enzymatic activities and TUNEL staining [Chan *et al.*,

Zhiguo Wang

2005; Corsten *et al.*, 2007]. Similarly, in MCF-7 human breast cancer cells, miR-21 also elicits anti-apoptotic effects [Si *et al.*, 2007; Zhu *et al.*, 2007]. miR-Lat was reported to protect neuroblastoma cells against apoptotic death [Gupta *et al.*, 2006].

On the other hand, several miRNAs have been referred to proapoptotic ones, which include let-7 family, miR-15a, miR-16-1, miR-29, miR-34a, miR-34b, miR-34c, miR-1, and miR-214. The best example of proapoptotic miRNAs is probably miR-15a and miR-16-1 that when forced to express induces apoptosis in chronic lymphocytic leukemia cells [Cimmino *et al.*, 2005]. Equally interesting is the finding that when Wi38 human diploid fibroblasts transduced with an AMO against miR-34 were treated with the apoptosis-inducing agent staurosporine, fewer early apoptotic cells and more viable cells were observed. The authors proposed that miR-34 can mediate key effects associated with p53 function for p53 transactivates gene expression of miR-34 [Raver-Shapira *et al.*, 2007]. Similar relationships between miR-34a and p53 have also been confirmed in other cells, such as H1299 human lung cancer cells, MCF-7 human breast cancer cells, U-2OS osteosarcoma cells [Tarasov *et al.*, 2007], and in p53 wild-type HCT116 colon cancer cells [Tazawa *et al.*, 2007]. Expression of miR-34a causes dramatic reprogramming of gene expression and promotes apoptosis [Chang *et al.*, 2007]. A most recent study further revealed that overexpression of miR-34a causes a dramatic reduction in cell proliferation through the induction of a caspase-dependent apoptotic pathway in three neuroblastoma cell lines Kelly, NGP and SK-N-AS [Welch *et al.*, 2007]. On the other hand, Bommer *et al* [2007] demonstrated that the expression of two miR-34b and miR-34c is dramatically reduced in 6 of 14 (43%) non-small cell lung cancers (NSCLCs) and that the restoration of miR-34 expression inhibits growth of NSCLC cells.

Nonetheless, precaution must be taken when attempting to categorize a miRNA by its role in apoptosis. Each miRNA has the potential to regulate >1000 protein-coding genes that could well be a mixture of some antiapoptotic and proapoptotic genes [Miranda *et al.*, 2006]. Whether a miRNA is antiapoptotic or proapoptotic may largely depend upon the cell-specific expression of genes involving in apoptosis and survival. There indeed have been a few instances reinforcing the needs to consider cell context as an important index for the role of miRNAs in apoptosis. As already mentioned above, miR-21 has been considered as antiapoptotic in glioblastoma [Chan *et al.*, 2005] and MCF-7 cells [Si *et al.*, 2007]. In HeLa cells, however, miR-21 does the opposite; inhibition of miR-21 increased the number of surviving cells [Cheng *et al.*, 2005]. Moreover, inhibition of miR-21 in A549 human lung cancer cells fails to alter cell death or growth [Cheng *et al.*, 2005]. Evidently, a same miRNA can have three different actions, antiapoptotic, proapoptotic or neutral, in different cell types. Similarly, miR-24 promotes growth in A549 cells, but inhibits growth in HeLa cells [Cheng *et al.*, 2005].

miRNAs AND CARDIOMYOCYTE APOPTOSIS

It is noticed that the previous work on miRNAs and apoptosis has been mostly limited to the context of cancer, while studies on apoptosis regulation by miRNAs in non-cancer cells have been sparse. The first evidence for the role of miRNAs in cardiomyocyte apoptosis was obtained in 2007 from my laboratory demonstrating the proapoptotic effect of miR-1 and anti-apoptotic effect of miR-133 in response to oxidative stress [Xu *et al.*, 2007]. Subsequent studies in 2009 and 2010 revealed the involvement of other miRNAs such as miR-21, miR-24 and miR-29 in ischemic myocardial injury [Yin *et al.*, 2008; Ye *et al.*, 2010].

Proapoptotic miRNAs

miR-1

Oxidative stress constitutes one of the major threats to the living system and accumulative oxidative damage has been implicated in many degenerative conditions including the aforementioned diseases such as heart failure, atherosclerosis and Alzheimer's disease [Ferrari *et al.*, 1998; Ide *et al.*, 2000; Nunomura *et al.*, 2001; Bennett, 2000; Markesbery, 1997], and even aging process, all of which are associated with progressive decrease in cell density. Indeed, oxidative stress is one of the most crucial factors causing neurodegenerative disorders. There is also a growing body of evidence that oxidative stress increases in myocardial failure and may contribute to the structural and functional changes that lead to disease progression [Ide *et al.*, 2000]. Moreover, oxidative stress is a well-known factor promoting apoptosis [Antunes *et al.*, 2001; Ren *et al.*, 2001; Han *et al.*, 2001; Wang *et al.*, 2002].

Mitochondrial death pathway is one of the major mechanisms for apoptosis induced by many cellular stresses including oxidative stress, which involves selective disruption of the outer membrane as a result of mitochondrial matrix hyperpolarization and/or matrix swelling, pore formation by proteins such as Bax and Bcl-xS, or rapid loss of $\Delta\Psi$ following permeability transition [Latchman, 2001; Gupta & Knowlton, 2005]. HSPs are expressed both constitutively (cognate proteins) and under stressful conditions (inducible forms), with constitutive expression being most prominent in mammalian tissues. HSPs are primarily anti-apoptotic and different HSPs have been shown to inhibit the mitochondrial death pathway at different points. HSP60 in the heart has key anti-apoptotic functions because of its ability to form complexes with Bax, Bak and Bcl-xS [Lin *et al.*, 2001; Shan *et al.*, 2003; Marber *et al.*, 1995], but not with Bcl-2. Binding of HSP60 in the normal cardiac cells prevents Bax from oligomerizing and inserting into the mitochondrial membrane. Reduction of HSP60 is associated with an overall decrease in Bcl-2 along with an increase in Bax and Bak and is sufficient to precipitate apoptosis [Lin *et al.*, 2001; Shan *et al.*, 2003; Marber *et al.*, 1995]. HSP70 exerts its antiapoptotic effect by preventing oligomerized Apaf-1 from recruiting pro-Casp9 [Latchman, 2001; Marber *et al.*, 1995]. HSP70 can also inhibit apoptosis in a caspase-independent manner by inhibiting the c-Jun N-terminal kinase (JNK kinase). However, Casp9 is a critical regulator of mitochondriamediated apoptosis; it forms a multimeric complex with cytochrome c and Apaf-1 to activate downstream caspases such as caspase-3 leading to apoptotic cell death [Han *et al.*, 2006; Kannan & Jain, 2000]. We reported that miR-1 and miR-133 produced opposing effects on apoptosis, induced by oxidative stress in H9c2 rat ventricular cells, with miR-1 being pro-apoptotic and miR-133 being antiapoptotic [Xu *et al.*, 2007].

We found that miR-1 level is significantly increased in response to oxidative stress in cardiomyocytes. This overexpression is involved in regulation of apoptotic cell death. Overexpression of miR-1 in H9c2 rat myoblasts provoked apoptotic cell death, which was partially rescued by treatment with miR-133 [Xu *et al.*, 2007]. Similar effects on oxidative stress–induced apoptosis were observed in H_2O_2-treated H9c2 cells, wherein cotransfection with miR-1 led to 60% reduction in the IC_{50} value necessary for oxidative stress–induced DNA fragmentation, and cotransfection with miR-133 resulted in a 40% increase in the IC_{50} value required for oxidative stress–induced DNA fragmentation. Qualitatively similar results were obtained when rat neonatal ventricular myocytes were exposed to H_2O_2. Subsequent computational analyses predicted that heat shock protein HSP60 and HSP70 were targets for miR-1 and that caspase-9 was a target for miR-133. This was verified experimentally by showing that miR-1 suppressed HSP60 and HSP70 protein levels by ~70% and ~60%, respectively, whereas miR-133 decreases protein levels of caspase-9 and caspase-9 activity levels in rat H9c2 cells. The post-transcriptional repression of HSP60 and HSP70 and caspase-9 was confirmed by luciferase reporter experiments. We postulated that the relative levels of miR-1 and miR-133 may determine cell fate.

It is noteworthy that miR-1 levels are overexpressed in coronary disease and are highly upregulated in acute myocardium ischemia (AMI) and in ischemia/reperfusion injury (I/R-I) [Yang *et al.*, 2007; Tang *et al.*, 2010; Yin *et al.*, 2008] but downregulated in cardiac hypertrophy/heart failure [Carè *et al.*, 2007; Sayed *et al.*, 2007; Yin *et al.*, 2008], wherein apoptosis is an important mode of cell death [Rodríguez *et al.*, 2002; Zidar *et al.*, 2007]. In this regard, it seems that increase in miR-1 during AMI and I/R-I contributes to apoptosis in these settings, and decrease in miR-1 in cardiac hypertrophy/heart failure is an adaptive change to minimize its deleterious effect. On the other hand, the cytoprotective effect of miR-133 may be weakened in cardiac hypertrophy/heart failure since miR-133 level has been found significantly reduced under such conditions and this may contribute to increased tendency of apoptosis induction in hypertrophic myocytes [Izumiya *et al.*, 2003; Tea *et al.*, 1999; Teiger *et al.*, 1996]. These notes merits future studies to verify.

The proapoptotic action of miR-1 in cardiomyocytes has been subsequently reproduced by other laboratories. A 2008 study by Yu *et al* [2008] investigated the possible miRNA mechanism for glucose cardiotoxicity. Glucose toxicity is an important initiator of cardiovascular disease, contributing to the development of cardiomyocyte death and diabetic complications. Yu *et al* [2008] showed that H9C2 cells exposed to high glucose have increased miR-1 expression level, decreased mitochondrial membrane potential, increased cytochrome-c release, and increased apoptosis. Glucose induced mitochondrial dysfunction, cytochrome-c release and apoptosis was blocked by IGF-1. miR-1 mimics, but not mutant miR-1, blocked the capacity of IGF-1 to prevent glucose-induced mitochondrial dysfunction, cytochrome-c release and apoptosis. The finding indicates that IGF-1 inhibits glucose-induced mitochondrial dysfunction, cytochrome-c release and apoptosis via downregulating miR-1 expression. The finding, according to the authors, provides a miRNA mechanism for the deleterious effects of glucose in the development of cardiomyocyte death and diabetic complications.

In a most recent study in 2010, Tang and colleagues studied the role of miR-1 in a rat model of ischemia/reperfusion injury (I/R-I). The level of miR-1 was found inversely correlated with Bcl-2 protein expression in cardiomyocytes with I/R-I. In vitro, the level of miR-1 was dramatically increased in response to H_2O_2. Overexpression of miR-1 facilitated H_2O_2-induced apoptosis in cardiomyocytes. Inhibition of miR-1 by antisense inhibitory oligonucleotides caused marked resistance to H_2O_2. Through bioinformatics, the authors identified the potential target sites for miR-1 on the 3'UTR of Bcl-2. miR-1 significantly reduced the expression of Bcl-2 at both mRNA and protein levels. The post-transcriptional repression of Bcl-2 was further confirmed by luciferase reporter experiments. The data indicate that miR-1 is not only proarrhythmic in AMI [Yang *et al.*, 2007] but also proapoptptic in I/R-I. The findings from these above studies also indicate that the proapoptotic action of miR-1 involves multiple death signaling pathways including HSP60/70, IGF-1 and Bcl-1, in a coordinated manner or separately under different situatrions.

miR-29

Pioglitazone, a peroxisome proliferator-activated receptor (PPAR)-γ agonist, has been documented by numerous studies to be able to limit myocardial infarct size in experimental animals [Yasuda *et al.*, 2009; Wynne *et al.*, 2005]. Ye *et al* [2010] assessed the effects of PPAR-γ activation on myocardial miRNA expression and the role of miRNAs in I/R-I in the rat heart after pioglitazone administration using miRNA microarray methods, followed by Northern Blot verification. They found that miR-29a and miR-29c levels were decreased after 7-day treatment with pioglitazone. In H9c2 cells, the effects of pioglitazone and rosiglitazone on miR-29 expression levels were blocked by a selective PPAR-γ inhibitor GW9662. Down-regulation of miR-29 by antisense inhibitor or by pioglitazone protected H9c2 cells from simulated IR injury, with increased cell survival and decreased caspase-3 activity. In contrast, overexpressing miR-29 promoted apoptosis and completely blocked the protective effect of pioglitazone. Antagomirs against miR-29a or -29c significantly reduced myocardial infarct size and apoptosis in hearts subjected to I/R-I. Western blot analyses demonstrated that Mcl-2, an anti-apoptotic Bcl-2 family member, was increased by miR-29 inhibition, similar to the finding in cancer cells [Mott *et al.*, 2007]. Clearly, down-regulation of miR-29 protected hearts against I/R-I through its anti-apoptotic activity. miR-29 thus represents another proapoptotic miRNA in cardiac cells, in addition to miR-1 [Xu *et al.*, 2007].

miR-320

Ren *et al* [2009] analyzed miRNA expression profile in murine hearts subjected to ischemia/reperfusion (I/R) *in vivo* and *ex vivo*, followed by qRT-PCR verification. They found that only miR-320 expression was significantly decreased in the hearts on I/R-I *in vivo* and *ex vivo*. Overexpression of miR-320 in cultured adult rat cardiomyocytes enhanced apoptotic cell death, whereas knockdown produced cytoprotective effect against apoptosis, on simulated I/R-I. Furthermore, transgenic mice with cardiac-specific overexpression of miR-320 revealed an increased extent of apoptosis and infarction size in the hearts on I/R-I *in vivo* and *ex vivo* relative to the wild-type controls. Conversely, *in vivo* treatment with antagomir-320 reduced infarction size relative to the administration of mutant antagomir-320 and saline controls. They subsequently identified heat-shock protein 20 (Hsp20), a cardioprotective protein, as a target for miR-320, utilizing a luciferase/GFP reporter activity assay and examining the expression of Hsp20 on miR-320 overexpression and knockdown in cardiomyocytes.

Antiapoptotic miRNAs

miR-133

As already described in the section for miR-1, overexpression of miR-133 acts against the proapoptotic effect of miR-1, upon co-transfection, in rat neonatal ventricular myocytes and in rat H9c2 cells with or without exposuse to H_2O_2 [Xu *et al.*, 2007]. We established that miR-133 represses protein level of caspase-9, which largely underlies its anti-apoptoic or cytoprotective action. We postulated that the relative levels of miR-1 and miR-133 may determine cell fate.

miR-21

In addition to miR-133, it appears that miR-21 is also an anti-apoptotic miRNA. Indeed, miR-21 has been found upregulated in various cancers (malignant glioblastoma, colorectal carcinoma, cervical adenocarcinoma) and, based on its in silico–predicted proapoptotic gene targets, has been proposed as a potential antiapoptotic miRNA [Si *et al.*, 2007; Asangani *et al.*, 2008; Frankel *et al.*, 2007]. Knockdown of miR-21 in glioblastoma cells leads to caspase activation and apoptotic cell death, and depletion of miR-21 in vascular smooth muscle cells leads to a dose-dependent increase in apoptosis and decrease in cell proliferation [Ji *et al.*, 2007]. As noted above, several independent groups have shown that the expression levels of miR-21 are increased 2- to 4-fold in hypertropied heart [Tatsuguchi *et al.*, 2007; Cheng *et al.*, 2007; Dong *et al.*, 2009]. Although miR-21 levels were not examined in relation to myocyte apoptosis in these studies, the upregulation of miR-21 may represent a prosurvival stress response in response to hemodynamic pressure overload.

Yin *et al* [2008] investigated the role of miRNA in protection against ischemia/reperfusion injury (I/R-I). Mice subjected to cytoprotective heat-shock (HS) showed a significant increase of miR-1 by ~80%, miR-21 by ~100% and miR-24 by ~60% in the heart, as determined by qPCR. miRNAs isolated from heat-shock mice and injected into non-HS mice significantly reduced infarct size after I/R-I, which was associated with the inhibition of pro-apoptotic genes and increase in anti-apoptotic genes. Chemically synthesized miR-21 also reduced infarct size, whereas a miR-21 inhibitor abolished this effect. These studies provide evidence for the potential role of endogenous miRNA in cardioprotection following I/R-I. miRNA treatment caused profound changes in several apoptotic related genes as determined by gene microarray analysis. The caspase family members 1, 2, 8 and 14 were suppressed in the hearts treated with miRNA from heat-shock mice as compared to the controls. Except for BNIP-3, most of the pro-apoptotic genes including Bid (BH3 interacting domain death agonist), Bcl-10 (B-cell leukemia/lymphoma 10), Cidea (cell death-inducing DNA fragmentation factor, alpha subunit-like effector A), Ltbr (lymphotoxin B receptor), Trp53 (transformation related protein 53), Fas (TNF receptor superfamily member) and Fasl (Fas ligand, TNF superfamily, member 6), were also repressed. On the other hand, the anti-apoptotic genes, Bag-3 (Bcl-2-associated athanogene and Prdx2 (Peroxiredoxin 2) were increased. According to previous studies, miR-1 has apoptosis-promoting effect [Xu *et al.*, 2007; Yu *et al.*, 2008; Tang *et al.*, 2010], miR-21 is anti-apoptotic in cardiac cells, and miR-24 is anti-apoptotic in cancer cells [Qin *et al.*, 2010] but direct evidence for its role in cardiomyocyte apoptiosis is yet to be provided. In particular, miR-1 was found to induce cardiomyocytes apoptosis in a rat model of I/R-I [Tang *et al.*, 2010]. It is possible that the anti-apoptotic effect of these miRNAs in I/R-I is a net outcome of the counteracting effects of these miRNAs. Moreover, most of the apoptosis-related proteins reported in this study are predicted not to be the targets for these miRNAs and whether the observed changes in their expression were ascribed to the upregulation of these miRNAs are not certain; by logic, the upregulation of the anti-apoptotic proteins could not be the direct consequence of the upregulation of the miRNAs.

In a subsequent study, Cheng *et al* [2009] confirmed the anti-apoptoptic effect of miR-21 in cardiac cells. Using quantitative real-time RT-PCR (qRT-PCR), the authors demonstrated that miR-21 was upregulated in cardiac myocytes after treatment with hydrogen peroxide (H_2O_2). H_2O_2-induced cardiac cell death and apoptosis were increased by miR-21 inhibitor and was decreased by pre-miR-21. Programmed cell death 4 (PDCD4) that was regulated by miR-21 and was established as a direct target of miR-21 in cardiac myocytes. Pre-miR-21-mediated protective effect on cardiac myocyte injury was inhibited in H_2O_2-treated cardiac cells via adenovirus-mediated overexpression of PDCD4 without miR-21 binding site. Moreover, Activator protein 1 (AP-1) was a downstream signaling molecule of PDCD4 that was involved in miR-21-mediated effect on cardiac myocytes. The results suggest that miR-21 is sensitive to H_2O_2 stimulation. miR-21 participates in H_2O_2-mediated gene regulation and functional modulation in cardiac myocytes. miR-21 might play an essential role in heart diseases related to oxidative stress such as cardiac hypertrophy, heart failure, myocardial infarction, and myocardial ischemia/reperfusion injury.

miR-199a

Hypoxia preconditioning of the heart is an immediate cellular reaction to brief hypoxia/reoxygenation cycles that stimulates *de novo* protein, but not mRNA, synthesis [Rowland *et al.*, 1997]. Hypoxia inducible factor (Hif)-1α is a transcription factor that is rapidly induced by hypoxia through a post-transcriptional mechanism, accounting for the transcription of 89% of genes that are upregulated during hypoxia [Wang *et al.*, 1993; Greijer *et al.*, 2005]. It has been well established that overexpression of Hif-1α during hypoxia resulted in a smaller infarct size following ischemia/reperfusion [Kido *et al.*, 2005]. Rane *et al* [2009] showed that downregulation of miR-199a during hypoxia is required for hypoxia-induced proapoptotic genes. This is because miR-199a directly targets Hif-1α mRNA and knockdown of miR-199a results in the upregulation of Hif-1α during normoxia. Forced expression of miR-199a during hypoxia inhibits Hif-1α expression and reduces apoptosis. Sirtuin (Sirt)1 is also a direct target of miR-199a and is responsible for downregulating prolyl hydroxylase 2, required for stabilization of Hif-1α.

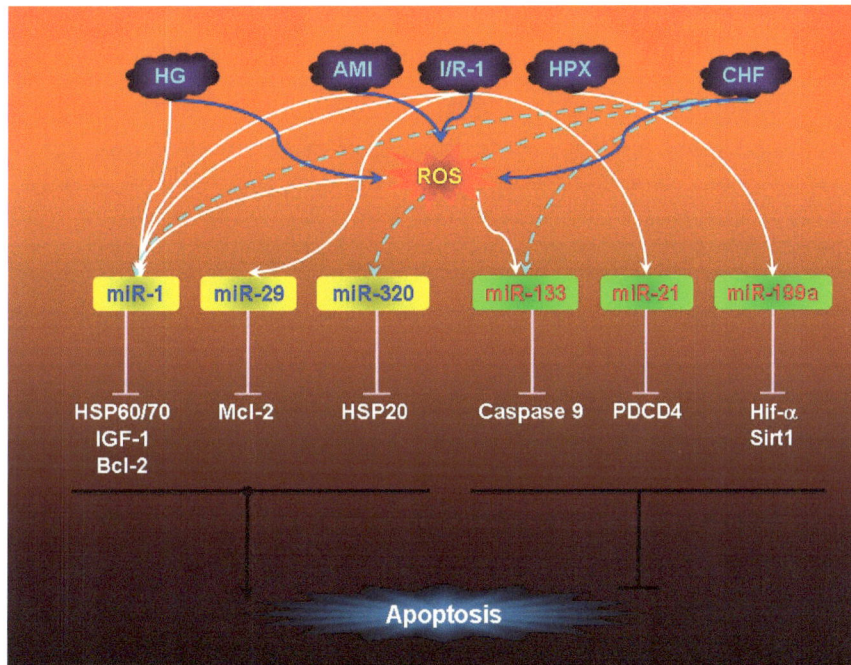

Figure 1: Diagram illustrating the role and signaling mechanisms of miRNAs in apoptosis of cardiac cells. Solid arrow indicates induction; dashed line indicates inhibition; and pause sign indicates repression. HG: high glucose; AMI: acute myocardial infarction; I/R-I: ischemia/reperfusion injury; ROS: reactive oxygen species.

In addition, a study described that treatment of mice with antagomiR-92a (anti-miRNA antisense conjugated with cholesterol moiety for *in vivo* application [Krützfeldt *et al.*, 2005] reduced apoptosis in the heart [Bonauer *et al.*, 2009]. However, the same antagomiR failed to affect apoptosis of cardiomyocytes *in vitro*, suggesting that the antiapoptotic activity of antagomiR-92a *in vivo* is mediated by an indirect mechanism.

miRNAs AND VASCULAR APOPTOSIS

It is known that proliferative vascular diseases share similar cellular events and molecular mechanisms with cancer, and neointimal lesion formation is the pathological basis of proliferative vascular diseases. Neointimal growth is the balance between proliferation and apoptosis of vascular smooth muscle cells (VSMCs). The increased VSMC proliferation or the relative decreased VSMC apoptosis are responsible for neointimal lesion formation. In a recent study of the potential roles of miRNAs in VSMCs proliferation and apoptosis [Ji *et al.*, 2007], the authors found that miR-21 level is up-regulated in proliferative VSMCs and depletion of miR-21 results in decreased cell proliferation and increased cell apoptosis in a dose-dependent manner in cultured rat aortic VSMCs. This suggests that miR-21 has a proproliferative and anti-apoptotic effect in VSMCs.Their preliminary data suggest that PTEN (phosphatase and tensin homology deleted on chromosome 10) and Bcl-2 might be involved in the anti-apoptotic action of miR-21 in VAMCs. The authors believe that miRNAs may be a new therapeutic target for proliferative vascular diseases such as atherosclerosis, postangioplasty restenosis, transplantation arteriopathy, and stroke.

One of the major roles miRNAs play is the ability of these molecules to control cell death that bears a wide range of biological and pathological implications. While this issue has been firmly established and well advanced in the field of oncology with a realistic hope in developing novel diagnostic/prognostic biomarkers and therapeutic agents as well, relatively little is yet known about the regulation of cardiac and vascular apoptosis by miRNAs and the role of the regulation in cardiovascular patholgenesis. Nonetheless, there are many reasons to believe that miRNAs are much more importantly involved in apoptosis in the cardiovascular system than we currently know; future studies will prove it. We are now just beginning to understand their role as gatekeepers of cell death.

REFERENCES

Antunes F, Cadenas E, Brunk UT. (2001) Apoptosis induced by exposure to a low steady-state concentration of H_2O_2 is a consequence of lysosomal rupture. Biochem J 365:549–555.

Asangani IA, Rasheed SA, Nikolova DA, Leupold JH, Colburn NH, Post S, Allgayer H. (2008) MicroRNA-21 (miR-21) post-transcriptionally downregulates tumor suppressor Pdcd4 and stimulates invasion, intravasation and metastasis in colorectal cancer. Oncogene 27:2128–2136.

Bonauer A, Carmona G, Iwasaki M, Mione M, Koyanagi M, Fischer A, Burchfield J, Fox H, Doebele C, Ohtani K, Chavakis E, Potente M, Tjwa M, Urbich C, Zeiher AM, Dimmeler S.(2009) MicroRNA-92a controls angiogenesis and functional recovery of ischemic tissues in mice. Science 324:1710–1713.

Carè A, Catalucci D, Felicetti F, Bonci D, Addario A, Gallo P, Bang ML, Segnalini P, Gu Y, Dalton ND, Elia L, Latronico MV, Høydal M, Autore C, Russo MA, Dorn GW, Ellingsen O, Ruiz-Lozano P, Peterson KL, Croce CM, Peschle C, Condorelli G. (2007) MicroRNA-133 controls cardiac hypertrophy. Nat Med 13:613–618.

Chan JA, Krichevsky AM, Kosik KS (2005) MicroRNA-21 is an antiapoptotic factor in human glioblastoma cells. Cancer Res 65:6029–6033.

Chen JF, Mandel EM, Thomson JM, Wu Q, Callis TE, Hammond SM, Conlon FL, Wang DZ. (2006) The role of microRNA-1 and microRNA-133 in skeletal muscle proliferation and differentiation. Nat Genet 38:228–233.

Chen JF, Murchison EP, Tang R, Callis TE, Tatsuguchi M, Deng Z, Rojas M, Hammond SM, Schneider MD, Selzman CH, Meissner G, Patterson C, Hannon GJ, Wang DZ. (2008) Targeted deletion of Dicer in the heart leads to dilated cardiomyopathy and heart failure. Proc Natl Acad Sci USA 105:2111–2116.

Cheng Y, Liu X, Zhang S, Lin Y, Yang J, Zhang C. (2009) MicroRNA-21 protects against the H_2O_2-induced injury on cardiac myocytes via its target gene PDCD4. J Mol Cell Cardiol 47:5–14.

Cheng Y, Zhu P, Yang J, Liu X, Dong S, Wang X, Chun B, Zhuang J, Zhang C. (2010). Ischemic preconditioning-regulated miR-21 protects the heart from ischemia/reperfusion injury via anti-apoptosis through its target PDCD4. Cardiovasc Res 2010 Mar 10.

Dong S, Cheng Y, Yang J, Li J, Liu X, Wang X, Wang D, Krall TJ, Delphin ES, Zhang C. (2009) MicroRNA expression signature and the role of microRNA-21 in the early phase of acute myocardial infarction. J Biol Chem 284:29514–29525.

Ferrari R, Agnoletti L, Comini L, Gaia G, Bachetti T, Cargnoni A, Ceconi C, Curello S, Visioli O. (1998) Oxidative stress during myocardial ischemia and heart failure. Eur Heart J 19:B2–B11.

Frankel LB, Christoffersen NR, Jacobsen A, Lindow M, Krogh A, Lund AH. (2007) Programmed cell death 4 (PDCD4) is an important functional target of the microRNA miR-21 in breast cancer cells. J Biol Chem 263:1026–1033.

Greijer AE, van der Groep P, Kemming D, Shvarts A, Semenza GL, Meijer GA, van de Wiel MA, Belien JA, van Diest PJ, van der Wall E. (2005) Up-regulation of gene expression by hypoxia is mediated predominantly by hypoxia-inducible factor 1 (HIF-1). J Pathol 206:291–304.

Gupta S, Knowlton AA. (2005) HSP60, Bax, apoptosis and the heart. J Cell Mol Med 9: 51–58.

Han H, Long H, Wang H, Wang J, Zhang Y, Wang Z (2004) Progressive apoptotic cell death triggered by transient oxidative insult in H9c2 rat ventricular cells: a novel pattern of apoptosis and the mechanisms. Am J Physiol 286:H2169–H2182.

Han H, Wang H, Long H, Nattel S, Wang Z. (2001) Oxidative preconditioning and apoptosis in L-cells: Roles of protein kinase B and mitogen-activated protein kinases. J Biol Chem 276:26357-26364.

Han Y, Chen YS, Liu Z, Bodyak N, Rigor D, Bisping E, Pu WT, Kang PM. (2006) Overexpression of HAX-1 protects cardiac myocytes from apoptosis through caspase-9 inhibition. Circ Res 99:415–423.

Ide T, Tsutsui H, Kinugawa S, Suematsu N, Hayashidani S, Ichikawa K, Utsumi H, Machida Y, Egashira K, Takeshita A. (2000) Direct evidence for increased hydroxyl radicals originating from superoxide in the failing myocardium. Circ Res 86:152–157.

Izumiya Y, Kim S, Izumi Y, Yoshida K, Yoshiyama M, Matsuzawa A, Ichijo H, Iwao H. (2003) Apoptosis signal-regulating kinase 1 plays a pivotal role in angiotensin II-induced cardiac hypertrophy and remodeling. Circ Res 93:874–883.

Ji R, Cheng Y, Yue J, Yang J, Liu X, Chen H, Dean DB, Zhang C. (2007) MicroRNA expression signature and antisense-mediated depletion reveal an essential role of microRNA in vascular neointimal lesion formation. Circ Res 100:1579–1588.

Kannan K, Jain SK. (2000) Oxidative stress and apoptosis. Pathophysiology 7:153–163.

Kido M, Du L, Sullivan CC, Li X, Deutsch R, Jamieson SW, Thistlethwaite PA. (2005) Hypoxia-inducible factor 1-alpha reduces infarction and attenuates progression of cardiac dysfunction after myocardial infarction in the mouse. J Am Coll Cardiol. 46:2116–2124.

Krützfeldt J, Rajewsky N, Braich R, Rajeev KG, Tuschl T, Manoharan M, Stoffel M. (2005) Silencing of microRNAs *in vivo* with 'antagomirs'. Nature 438:685–689.

Latchman DS. (2001) Heat shock proteins and cardiac protection. Cardiovasc Res 51:637–646.

Lee Y, Ahn C, Han J, Choi H, Kim J, Yim J, Lee J, Provost P, Rådmark O, Kim S, Kim VN. (2003) The nuclear RNase III Drosha initiates microRNA processing. Nature 425:415–419.

Lee Y, Hur I, Park SY, Kim YK, Suh MR, Kim VN. (2006) The role of PACT in the RNA silencing pathway. EMBO J 25:522–532.

Lin KM, Lin B, Lian IY, Mestril R, Scheffler I, Dillmann WH. (2001) Combined and individual mitochondrial HSP60 and HSP10 expression in cardiac myocytes protects mitochondrial function and prevents apoptotic cell deaths induced by simulated ischemia-reoxygenation. Circulation 103:1787–1792.

Lin H, Xiao J, Luo X, Xu C, Gao H, Wang H, Yang B, Wang Z. (2007) Overexpression HERG K+ channel gene mediates cell-growth signals on activation of oncoproteins Sp1 and NF-□B and inactivation of tumor suppressor Nkx3.1. J Cell Physiol 212:137–147.

Lu Y, Xiao J, Lin H, Bai Y, Luo X, Wang Z, Yang B. (2009) Complex antisense inhibitors offer a superior approach for microRNA research and therapy. Nucleic Acids Res 37:e24–e33.

Marber MS, Mestril R, Chi SH, Sayen MR. (1995) Overexpression of the rat inducible 70 kDa heat shock protein in a transgenic mouse increases the resistance of the heart to ischemic injury. J Clin Invest 95:1446–1456.

Markesbery WR. (1997) Oxidative stress hypothesis in Alzheimer's disease. Free Radic Biol Med 23:134–147.

Mott JL, Kobayashi S, Bronk SF, Gores GJ. (2007) mir-29 regulates Mcl-1 protein expression and apoptosis. Oncogene 26:6133–6140.

Nunomura A, Perry G, Aliev G, Hirai K, Takeda A, Balraj EK, Jones PK, Ghanbari H, Wataya T, Shimohama S, Chiba S, Atwood CS, Petersen RB, Smith MA. (2001) Oxidative damage is the earliest event in Alzheimer disease. J Neuropathol Exp Neurol 60:759–767.

Qin W, Shi Y, Zhao B, Yao C, Jin L, Ma J, Jin Y. (2010) miR-24 regulates apoptosis by targeting the open reading frame (ORF) region of FAF1 in cancer cells. PLoS One 5:e9429.

Rane S, He M, Sayed D, Vashistha H, Malhotra A, Sadoshima J, Vatner DE, Vatner SF, Abdellatif M. (2009) Downregulation of miR-199a derepresses hypoxia-inducible factor-1alpha and Sirtuin 1 and recapitulates hypoxia preconditioning in cardiac myocytes. Circ Res 104:879–886.

Ren JG, Xia HL, Just T, Dai YR. (2001) Hydroxyl radical-induced apoptosis in human tumor cells is associated with telomere shortening but not telomerase inhibition and caspase activation. FEBS Lett 488:123–132.

Ren XP, Wu J, Wang X, Sartor MA, Qian J, Jones K, Nicolaou P, Pritchard TJ, Fan GC. (2009) MicroRNA-320 is involved in the regulation of cardiac ischemia/reperfusion injury by targeting heat-shock protein 20. Circulation 119:2357–2366.

Rodríguez M, Lucchesi BR, Schaper J. (2002) Apoptosis in myocardial infarction. Ann Med 34:470–9.

Rowland RT, Meng X, Cleveland JC, Meldrum DR, Harken AH, Brown JM. (1997) Cardioadaptation induced by cyclic ischemic preconditioning is mediated by translational regulation of de novo protein synthesis. J Surg Res 71:155–160.

Sayed D, Hong C, Chen IY, Lypowy J, Abdellatif M. (2007) MicroRNAs play an essential role in the development of cardiac hypertrophy. Circ Res 100:416–424.

Shan YX, Liu TJ, Su HF, Samsamshariat A, Mestril R, Wang PH. (2003) Hsp10 and Hsp60 modulate Bcl-2 family and mitochondria apoptosis signaling induced by doxorubicin in cardiac muscle cells. J Mol Cell Cardiol 35:1135–1143.

Si ML, Zhu S, Wu H, Lu Z, Wu F, Mo YY. (2007) miR-21-mediated tumor growth. Oncogene 26:2799–1803.

Tang Y, Zheng J, Sun Y, Wu Z, Liu Z, Huang G. (2009) MicroRNA-1 regulates cardiomyocyte apoptosis by targeting Bcl-2. Int Heart J 50:377–387.

Tatsuguchi M, Seok HY, Callis TE, Thomson JM, Chen JF, Newman M, Rojas M, Hammond SM, Wang DZ. (2007) Expression of microRNAs is dynamically regulated during cardiomyocyte hypertrophy. J Mol Cell Cardiol 42:1137–1141.

Tea BS, Dam TV, Moreau P, Hamet P, deBlois D. (1999) Apoptosis during regression of cardiac hypertrophy in spontaneously hypertensive rats. Temporal regulation and spatial heterogeneity. Hypertension 34:229–235.

Teiger E, Than VD, Richard L, Wisnewsky C, Tea BS, Gaboury L, Tremblay J, Schwartz K, Hamet P. (1996) Apoptosis in pressure overload-induced heart hypertrophy in the rat. J Clin Invest 97:2891–2897.

Wang GL, Semenza GL. (1993) General involvement of hypoxia-inducible factor 1 in transcriptional response to hypoxia. Proc Natl Acad Sci USA 90:4304–4308.

Wang H, Zhang Y, Han H, Cao L, Wang J, Long H, Nattel S, Wang Z. (2002) HERG K$^+$ channel: A regulator of tumor cell apoptosis and proliferation. Cancer Res 62:4843–4848.

Wang Z, Luo X, Lu Y, Yang B. (2008) miRNAs at the heart of the matter. J Mol Med 86:772–783.

Weiler J, Hunziker J, Hall J. (2006) Anti-miRNA oligonucleotides (AMOs): ammunition to target miRNAs implicated in human disease? Gene Ther 13:496–502.

Wu H, Lima WF, Zhang H, Fan A, Sun H, Crooke ST. (2004) Determination of the role of the human RNase H1 in the pharmacology of DNA-like antisense drugs. J Biol Chem 279:17181–17189.

Wynne AM, Mocanu MM, Yellon DM. (2005) Pioglitazone mimics preconditioning in the isolated perfused rat heart: a role for the prosurvival kinases PI3K and P42/44MAPK. J Cardiovasc Pharmacol 46:817–822.

Xu C, Lu Y, Pan Z, Chu W, Luo X, Lin H, Xiao J, Shan H, Wang Z, Yang B. (2007) The muscle-specific microRNAs miR-1 and miR-133 produce opposing effects on apoptosis by targeting HSP60, HSP70 and caspase-9 in cardiomyocytes. J Cell Sci 120:3045–3052.

Yang B, Lin H, Xiao J, Lu Y, Luo X, Li B, Zhang Y, Xu C, Bai Y, Wang H, Chen G, Wang Z. (2007) The muscle-specific microRNA miR-1 causes cardiac arrhythmias by targeting GJA1 and KCNJ2 genes. Nat Med 13:486–491.

Yang B, Lu Y, Wang Z. (2009) MicroRNAs and apoptosis: implications for molecular therapy of human disease. Clin Exp Pharmacol Physiol 36:951–960.

Yasuda S, Kobayashi H, Iwasa M, Kawamura I, Sumi S, Narentuoya B, Yamaki T, Ushikoshi H, Nishigaki K, Nagashima K, Takemura G, Fujiwara T, Fujiwara H, Minatoguchi S. (2009) Antidiabetic drug pioglitazone protects the heart via activation of PPAR-gamma receptors, PI3-kinase, Akt, and eNOS pathway in a rabbit model of myocardial infarction. Am J Physiol Heart Circ Physiol 2009;296:H1558-1565.

Ye Y, Hu Z, Lin Y, Zhang C, Perez-Polo JR. (2010) Down-regulation of microRNA-29 by antisense inhibitors and a PPAR-{gamma} agonist protects against myocardial ischemia-reperfusion injury. Cardiovasc Res. 2010 Feb 17. [Epub ahead of print]

Yin C, Wang X, Kukreja RC (2008) Endogenous microRNAs induced by heat-shock reduce myocardial infarction following ischemia-reperfusion in mice. FEBS Lett 582:4137–4142.

Yin C, Salloum FN, Kukreja RC. (2009) A novel role of microRNA in late preconditioning: upregulation of endothelial nitric oxide synthase and heat shock protein 70. Circ Res 104:572–575.

Yu XY, Song YH, Geng YJ, Lin QX, Shan ZX, Lin SG, Li Y. (2008) Glucose induces apoptosis of cardiomyocytes via microRNA-1 and IGF-1.Biochem Biophys Res Commun 376:548–552.

Zidar N, Jera J, Maja J, Dusan S. (2007) Caspases in myocardial infarction. Adv. Clin. Chem 44: 1–33.

Zimmermann TS, Lee AC, Akinc A, Bramlage B, Bumcrot D, Fedoruk MN, Harborth J, Heyes JA, Jeffs LB, John M, Judge AD, Lam K, McClintock K, Nechev LV, Palmer LR, Racie T, Röhl I, Seiffert S, Shanmugam S, Sood V, Soutschek J, Toudjarska I, Wheat AJ, Yaworski E, Zedalis W, Koteliansky V, Manoharan M, Vornlocher HP, MacLachlan I. (2006) RNAi-mediated gene silencing in non-human primates. Nature 441:111–114.

miRNAs in Cardiac Contraction

Abstract: This chapter aims to introduce the role of miRNAs in regulating cardiac contraction. Cardiac contraction is triggered by excitation–contraction coupling: the cascade of biological events that begins with cardiac action potential and ends with myocyte contraction and relaxation. Cardiac muscle contraction is determined by the intrinsic contractile proteins: α- and β-myosin heavy chain (αMHC and βMHC). αMHC and βMHC are encoded by MYH6 and MYH7 genes, respectively, and their expression is species specific and varies in response to developmental and pathophysiological signaling alterations. Remarkably, studies revealed that myosin genes not only encode the major contractile proteins of muscle, but also act more broadly to control muscle gene expression and performance through a network of intronic miRNAs: the transcripts from these genes all contain pre-miRNAs. On the other hand, the cytoskeleton of cardiac myocytes consists of actin, the intermediate filament desmin, the sarcomeric protein titin, and α- and β-tubulin, which form the microtubules by polymerization. The loss of integrity of the cytoskeleton, with a resultant loss of linkage of the sarcomere to the sarcolemma and extracellular matrix, would be expected to lead to contractile dysfunction. miRNAs have been found to regulate both the contractile proteins (miR-208 and miR-21) and cytoskeleton proteins (miR-1 and miR-133) to regulate cardiac contraction.

INTRODUCTION

Cardiac muscle contraction is determined by the intrinsic contractile proteins: α- and β-myosin heavy chain (αMHC and βMHC, respectively). They are the major contractile proteins of cardiac muscles and the ratio of αMHC to βMHC expression is the primary determinant of the speed and force of contraction. αMHC and βMHC are encoded by MYH6 and MYH7 genes, respectively [Weiss & Leinwand, 1996], and their expression is species specific and varies in response to developmental and pathophysiological signaling alterations.

Cardiac contraction is triggered by excitation–contraction coupling: the cascade of biological events that begins with cardiac action potential and ends with myocyte contraction and relaxation. Classic studies from explanted failing hearts have shown that patients with end-stage heart failure exhibit decreased contractility and impaired relaxation, which is believed to be secondary to changes in the abundance and/or phosphorylation state of critical calcium (Ca^{2+}) regulatory proteins that are thought to play an important role in cross-bridge activation and relaxation. An additional defect in myocyte function in hypertrophy and heart failure is thought to occur secondary to changes in the actin and myosin myofibrillar cross-bridges.

In adult human hearts, βMHC is the predominant isoform and becomes more so during cardiac disease when αMHC is downregulated [Lowes *et al.*, 1997; Miyata *et al.*, 2000]. Mechanical stress induces a shift from the fast myosin αMHC to the slow myosin βMHC composition in the adult heart [Fatkin *et al.*, 2000; Gupta, 2007; Krenz & Robbins, 2004], which correlates with a decline in mechanical performance and contractile efficiency [Lowes *et al.*, 1997; Miyata *et al.*, 2000]. Indeed, early studies showed that myofibrillar ATPase was reduced in the hearts of patients who died of heart failure and that these abnormalities in ATPase activity could be explained by an isoform switch from αMHC, which hydrolyzes ATP rapidly and is expressed in the adult heart, to βMHC, which hydrolyzes more slowly and is expressed in the fetal heart. Whereas αMHC accounts for approximately 33% of MHC mRNA in normal human myocardium, the abundance of αMHC mRNA decreases to 2% in the failing heart [Nakao *et al.*, 1997]. Relatively minor changes in the ratio of these two myosin isoforms can have profound effects on cardiac contractility in human and rodent hearts [Herron & McDonald, 2002; Schiaffino & Reggiani, 1996], and increased expression of βMHC in the myocardium decreases power output and contributes to depressed systolic function in end-stage heart failure. In addition to MYH6 and MYH7 genes, another gene MYH7b also encodes a myosin heavy chain protein (βMHC7b).

miRNAs AND MYOSIN HEAVY CHAIN PROTEINS

Remarkably, studies revealed that myosin genes not only encode the major contractile proteins of muscle, but also act more broadly to control muscle gene expression and performance through a network of intronic miRNAs: the transcripts from these genes all contain pre-miRNAs [Berezikov *et al.*, 2006; Landgraf *et al.*, 2007] that can be processed to become mature miRNAs. Earlier in 2007, Olson's group identified a novel miRNA miR-208a from MYH6 gene that is specifically expressed in cardiac muscles [van Rooij *et al.*, 2007]. Subsequently in 2009, the same group further characterized another two intronic miRNAs (miR-208b and miR-499) generated from MYH7 and MYH7b genes that are expressed in both cardiac and slow skeletal muscles.

miR-208a

One of the best studied miRNAs linked to cardiac contraction is miR-208a. Studies have shown that miR-208 is upregulated in response to a hemodynamic pressure overload, as well as in human heart failure [Tatsuguchi *et al.*, 2007; Thum *et al.*, 2007]. Olson and colleagues [van Rooij *et al.*, 2007] showed that miR-208a, a cardiac-specific miRNA, is coordinately regulated with αMHC. Mice deficient for miR-208a appeared phenotypically normal at baseline; however, these mice had a blunted hypertrophic response following pressure overload by TAB, as well as decreased myocardial fibrosis. Of note, miR-208a-deficient mice did not express αMHC in response to pressure overload. Although mice with cardiac-restricted overexpression of miR-208a had increased expression of αMHC, they did not develop pathological remodeling. These observations suggest that miR-208a is required for the development of cardiac hypertrophy and myocardial fibrosis and that miR-208 is a positive regulator of αMHC gene expression [van Rooij *et al.*, 2007]. Whereas in the adult heart, absence of miR-208a results in severe blunting of αMHC expression in response to pressure overload or activated calcineurin, expression of αMHC remains unaltered in hearts of newborn miR-208a–null mice, demonstrating that this miRNA specifically participates in the mechanism of stress-dependent regulation of αMHC [van Rooij *et al.*, 2007]. Thus, miR-208a regulates αMHC expression postnatally, acting as an on–off switch rather than a fine tuner of expression.

In addition, miR-208a was found to control the expression of βMHC in response to cardiac stress [van Rooij *et al.*, 2007]: miR-208a null mice subjected to pressure overload (surgically induced by aortic constriction) or pathological remodeling (signaled by calcineurin) did not experience the usual hypertrophic or fibrotic processes occurring in the heart with stress. While pressure overload strongly induced fetal genes in miR-208a-deficient mice, such as atrial and brain natriuretic peptides, it failed to affect the expression of βMHC. THRAP1, the co regulator of the thyroid hormone receptor, was identified as a target of miR-208a. In the absence of miR-208a, THRAP1 becomes upregulated, resulting in increased actions of this protein on the hormone response elements, and the βMHC promoter is consequently repressed by an enhanced negative hormone response element. This miRNA has been found upregulated with hypertrophy and heart failure [Thum *et al.*, 2007; Tatsuguchi *et al.*, 2007].

miR-21

Alterations in the expression or activity, or both, of myofilament regulatory proteins has been proposed as a potential mechanism for the decrease in cardiac contractile function in heart failure, including changes in myosin light chains and the troponin–tropomyosin complex. Studies have implicated a role for miR-21 in regulating hypertrophic growth in cardiac myocytes [Sayed *et al.*, 2007; Cheng *et al.*, 2007; Tatsuguchi *et al.*, 2007]. Although not yet studied in human heart failure, miR-21 targets the 3'UTR of tropomyosin and inhibits its translation in tumor cells [Zhu *et al.*, 2007]. Further studies will be necessary to delineate the role of miR-21 in cross-bridge cycling in heart failure.

miRNAs AND CYTOSKELETON

The cardiac myofilaments are composed of highly ordered arrays of proteins that coordinate cardiac contraction and relaxation in response to the rhythmic waves of $[Ca^{2+}]_i$ during the cardiac cycle. Several cardiac disease states are associated with altered myofilament protein interactions that contribute to cardiac dysfunction. The cytoskeleton as classically defined for eukaryotic cells consists of three systems of protein filaments: the microtubules, the intermediate filaments, and the microfilaments. In mature striated muscle such as the heart of the adult mammal, these three types of cytoskeletal filaments are superimposed spatially on the myofilaments, a specialized system of contractile protein filaments. Each of these systems of protein filaments has the potential to respond in an adaptive or maladaptive manner during load-induced hypertrophic cardiac growth. However, the extent to which such hypertrophy is compensatory is also critically dependent on the type of hemodynamic overload that serves as the hypertrophic stimulus. The loss of integrity of the cytoskeleton, with a resultant loss of linkage of the sarcomere to the sarcolemma and extracellular matrix, would be expected to lead to contractile dysfunction. The cytoskeleton of cardiac myocytes consists of actin, the intermediate filament desmin, the sarcomeric protein titin, and α- and β-tubulin, which form the microtubules by polymerization. Vinculin, talin, dystrophin, and spectrin represent a separate group of membrane-associated cytoskeletal proteins. Disruption of cytoskeletal and/or membrane-associated proteins has been implicated in the pathogenesis of heart failure in numerous studies [Latronico & Condorelli, 2009]. At present, there is limited evidence that miRNAs are involved in regulating the cytoskeleton; insofar as relatively few studies have examined these targets.

The decreased expression of miR-1 and miR-133 allows for the increased expression of growth-related genes that are responsible for cardiac hypertrophy [see Chapter 3]. Adenoviral-mediated overexpression of miR-1 suppressed sarcomeric α-actin organization that is normally observed in serum-deprived neonatal cardiac myocytes. The effects of miR-1 on cytoskeletal reorganization were also thought to be responsible for the inhibition of endothelin-induced cell spreading in neonatal myocytes. Similarly, adenoviral transfection with miR-133 inhibited the reorganization of the actin myofilaments observed in hypertrophic growth of cardiac myocytes [Carè *et al.*, 2007]. Two validated target genes of miR-133 are RhoA and Cdc42, the members of the Rho family of GTP binding proteins that are involved in myofibrillar rearrangement and are important for cell motility and contractility. miR-1 also targets a number of cytoskeletal-related proteins, including microtubule-related proteins, kinectins, actin binding proteins, and cadherins. Insofar as miR-1 and miR-133 do not appear to be downregulated in human heart failure, the significance of the aforementioned miR-induced changes in the cytoskeleton vis-a-vis the biology of the failing myocyte remains to be determined.

miRNAs AND EXCITATION–CONTRACTION COUPLING

As described in Chapter 6, alterations in intracellular Ca^{2+} cycling have been implicated in different cardiac diseases, including arrhythmia and heart failure. Cardiac contractility is modulated by SR Ca^{2+} release machinery, including phosphatases PP1 and PP2A catalytic subunit [Steenaart *et al.*, 1992; Davare *et al.*, 2000; Marx *et al.*, 2000], the latter of which has been shown to complex with dihydropyridine receptor (DHPR) and RyR2 and is critical to dephosphorylation of these proteins following their phosphorylation by PKA and/or CaMKII [Davare *et al.*, 2000; Marx *et al.*, 2000]. Consistent with the role of B56α in conveying PP2A phosphatase activity to DHPR, B56α has been shown to coimmunoprecipitate with the α–subunit of the cardiac L-type Ca^{2+} channel (Cav1.2α) [Hall *et al.*, 2006]. Terentye *et al* [2009] investigated the effects of increased expression of miR-1 on excitation– contraction coupling and Ca^{2+} cycling in rat ventricular myocytes using methods of electrophysiology, Ca^{2+} imaging and quantitative immunoblotting. Adenoviral-mediated overexpression of miR-1 in myocytes resulted in a marked increase in the amplitude of the inward Ca^{2+} current, flattening of Ca^{2+} transients voltage dependence, and enhanced frequency of spontaneous Ca^{2+} sparks while reducing the sarcoplasmic reticulum Ca^{2+} content as compared with control. miR-1 overexpression increased phosphorylation of the ryanodine receptor (RyR2). Overexpression of miR-1 was accompanied by a selective decrease in expression of the protein phosphatase PP2A regulatory subunit B56α involved in PP2A targeting to specialized subcellular domains. The authors concluded that miR-1 enhances cardiac excitation–contraction coupling by selectively increasing phosphorylation of the L-type and RyR2 channels via disrupting localization of PP2A activity to these channels. miR-1 enhanced the functional activity of RyR2 channels and thus resulted in increased EC coupling gain, elevated diastolic SR Ca^{2+} leak, and reduced SR Ca^{2+} content [Terentye *et al.*, 2009].

REFERENCES

Berezikov E, Thuemmler F, van Laake, LW, Kondova I, Bontrop R, Cuppen E, Plasterk RH. (2006) Diversity of microRNAs in human and chimpanzee brain. Nat Genet 38:1375–1377.

Carè A, Catalucci D, Felicetti F, Bonci D, Addario A, Gallo P, Bang ML, Segnalini P, Gu Y, Dalton ND, Elia L, Latronico MV, Høydal M, Autore C, Russo MA, Dorn GW, Ellingsen O, Ruiz-Lozano P, Peterson KL, Croce CM, Peschle C, Condorelli G. (2007) MicroRNA-133 controls cardiac hypertrophy. Nat Med 13:613–618.

Davare MA, Horne MC, Hell JW. (2000) Protein phosphatase 2A is associated with class C L-type calcium channels (Cav1.2) and antagonizes channel phosphorylation by cAMP-dependent protein kinase. J Biol Chem 275:39710–39717.

Fatkin D, McConnell BK, Mudd JO, Semsarian C, Moskowitz IG, Schoen FJ, Giewat M, Seidman CE, Seidman JG. (2000) An abnormal Ca^{2+} response in mutant sarcomere protein-mediated familial hypertrophic cardiomyopathy. J Clin Invest 106:1351–1359.

Gupta MP. (2007) Factors controlling cardiac myosin-isoform shift during hypertrophy and heart failure. J Mol Cell Cardiol 43:388–403.

Hall DD, Feekes JA, Arachchige Don AS, Shi M, Hamid J, Chen L, Strack S, Zamponi GW, Horne MC, Hell JW. (2006) Binding of protein phosphatase 2A to the L-type calcium channel Cav1.2 next to Ser1928, its main PKA site, is critical for Ser1928 dephosphorylation. Biochemistry 45:3448–3459.

Herron TJ, McDonald KS. (2002) Small amounts of alpha-myosin heavy chain isoform expression significantly increase power output of rat cardiac myocyte fragments. Circ Res 90:1150–1152.

Krenz M, Robbins J. (2004) Impact of beta-myosin heavy chain expression on cardiac function during stress. J Am Coll Cardiol 44:2390–2397.

Latronico MV, Condorelli G. (2009) MicroRNAs and cardiac pathology. Nat Rev Cardiol 6:419–429.

Landgraf P, Rusu M, Sheridan R, Sewer A, Iovino N, Aravin A, Pfeffer S, Rice A, Kamphorst AO, Landthaler M, Lin C, Socci ND, Hermida L, Fulci V, Chiaretti S, Foà R, Schliwka J, Fuchs U, Novosel A, Müller RU, Schermer B, Bissels U, Inman J, Phan Q, Chien M, Weir DB, Choksi R, De Vita G, Frezzetti D, Trompeter HI, Hornung V, Teng G, Hartmann G, Palkovits M, Di Lauro R, Wernet P, Macino G, Rogler CE, Nagle JW, Ju J, Papavasiliou FN, Benzing T, Lichter P, Tam W, Brownstein MJ, Bosio A, Borkhardt A, Russo JJ, Sander C, Zavolan M, Tuschl T. (2007) A mammalian microRNA expression atlas based on small RNA library sequencing. Cell 129:1401–1414.

Lowes BD, Minobe W, Abraham WT, Rizeq MN, Bohlmeyer TJ, Quaife RA, Roden RL, Dutcher DL, Robertson AD, Voelkel NF, Badesch DB, Groves BM, Gilbert EM, Bristow MR. (1997) Changes in gene expression in the intact human heart. Downregulation of alpha-myosin heavy chain in hypertrophied, failing ventricular myocardium. J Clin Invest 100:2315–2324.

Marx SO, Reiken S, Hisamatsu Y, Jayaraman T, Burkhoff D, Rosemblit N, Marks AR. (2000) PKA phosphorylation dissociates FKBP12.6 from the calcium release channel (ryanodine receptor): defective regulation in failing hearts. Cell 101:365–376.

Miyata S, Minobe W, Bristow MR, Leinwand LA. (2000) Myosin heavy chain isoform expression in the failing and nonfailing human heart. Circ Res 86:386–390.

Morkin E. (2000) Control of cardiac myosin heavy chain gene expression. Microsc Res Tech 50:522–531.

Nakao K, Minobe W, Roden R, Bristow MR, Leinwand LA. (1997) Myosin heavy chain gene expression in human heart failure. J Clin Invest 100:2362–2370.

Schiaffino S, Reggiani C. (1996) Molecular diversity of myofibrillar proteins: gene regulation and functional significance. Physiol Rev 76:371–423.

Steenaart NAE, Ganim JR, DiSalvo J, Kranias EG. (1992) The phospholamban phosphatase associated with cardiac sarcoplasmic reticulum is a type 1 enzyme. Arch Biochem Biophys 293:17–24.

Tatsuguchi M, Seok HY, Callis TE, Thomson JM, Chen JF, Newman M, Rojas M, Hammond SM, Wang DZ. (2007) Expression of microRNAs is dynamically regulated during cardiomyocyte hypertrophy. J Mol Cell Cardiol 42:1137–1141.

Terentyev D, Belevych AE, Terentyeva R, Martin MM, Malana GE, Kuhn DE, Abdellatif M, Feldman DS, Elton TS, Györke S. (2009) miR-1 overexpression enhances Ca^{2+} release and promotes cardiac arrhythmogenesis by targeting PP2A regulatory subunit B56alpha and causing CaMKII-dependent hyperphosphorylation of RyR2. Circ Res 104:514–521.

Thum T, Galuppo P, Wolf C, Fiedler J, Kneitz S, vanLaake LW, Doevendans PA, Mummery CL, Borlak J, Haverich A, Gross C, Engelhardt S, Ertl G, Bauersachs J. (2007) MicroRNAs in the human heart: a clue to fetal gene reprogramming in heart failure. Circulation 116:258–267.

van Rooij E, Quiat D, Johnson BA, Sutherland LB, Qi X, Richardson JA, Kelm RJ Jr, Olson EN. (2009) A family of microRNAs encoded by myosin genes governs myosin expression and muscle performance. Dev Cell 17:662–673.

van Rooij E, Sutherland LB, Qi X, Richardson JA, Hill J, Olson EN. (2007) Control of stress-dependent cardiac growth and gene expression by a microRNA. Science 316:575–579.

Weiss A, Leinwand LA. (1996) The mammalian myosin heavy chain gene family. Annu Rev Cell Dev Biol 12:417–439.

Zhu SM, Si ML, Wu HL, Mo YY. (2007) MicroRNA-21 targets the tumor suppressor gene tropomyosin 1 (TPM1). J Biol Chem 282:14328–14336.

CHAPTER 15

miRNAs in Neurohormonal Activation

Abstract: This chapter aims to introduce the role of miRNAs in regulating neurohormonal activation. The natural progression of heart failure is accompanied by the compensatory activation of cardiac and extracardiac neurohormonal systems and changes in the anatomy and function of the left ventricle. An array of biologically active molecules belong to the sympathetic adrenergic nervous system (norepinephrine) and renin–angiotensin–aldosterone system (RAS) (Ang II and aldosterone), which are responsible for maintaining cardiac output through increased retention of salt and water, peripheral arterial vasoconstriction, and contractility, as well as inflammatory mediators that are responsible for cardiac repair and remodeling. Although, the role of miRNAs in regulating the components of RAS and the adrenergic system is still poorly not well understood, several recent observations are worth noting. In particular, miR-155 is implicated in suppressing the levels of the Ang II type 1 receptor, and miR-21 can increase aldosterone secretion in human adrenal cells. This chapter describes very limited information in this regard.

INTRODUCTION

Activation of neurohormonal systems resulting in expression of biologically active molecules (collectively referred to as neurohormones) is essential for homeostasis in the normal heart, but it also plays an important role in the progression of heart failure. In acute heart failure, neurohormonal activation has beneficial effects, but in chronic heart failure their activation produces deleterious effects by increasing the load on the left ventricle and promoting structural remodeling, which may further impair left ventricular function [Mann & Bristow, 2005]. Thus, neurohormonal activation is a commonly cited array of phenomena in the body's physiologic response to heart failure. In adult congenital heart disease, neurohormonal activation characterizes chronic heart failure, relating to symptom severity and ventricular dysfunction. The portfolio of the known neurohormones include (1) the sympathetic adrenergic nervous system (norepinephrine), (2) renin–angiotensin–aldosterone system (RAS) (Ang II and aldosterone), (3) vasopressinergic system, (4) natriuretic peptide systems, which are responsible for maintaining cardiac output through increased retention of salt and water, peripheral arterial vasoconstriction, and contractility, and (5) inflammatory mediators (tumor necrosis factor, interleukin-1, and interleukin-6) that are responsible for cardiac repair and remodeling [Divakaran & Mann, 2008].

Recent studies have demonstrated that treating isolated neonatal rat ventricular myocytes with ligands that mimic neurohormonal activation, such as PE, endothelin-1, or Ang II is sufficient to upregulate miRNAs including miR-21, miR-23a, and miR-133, for hypertrophic growth [Tatsuguchi *et al.*, 2007; Cheng *et al.*, 2007; Sayed *et al.*, 2007]. The involvement of miRNAs in "neurohormonal activation" through enhanced RAS signaling has also been reported.

miR-21

Angiotensin II is an important modulator of adrenal zona glomerulosa cell physiology and heart failure. Romero *et al* [2008] studied the involvement of miRNAs in angiotensin II-mediated adrenocortical cell physiology. They screened for miRNAs regulated by angiotensin II in the human adrenocortical cell line H295R and found that miR-21 expression levels were specifically modulated by angiotensin II. Angiotensin II increased miR-21 expression in a time-dependent fashion, reaching a 4.4-fold induction after 24 h stimulation. Angiotensin II-mediated miR-21 expression resulted in biologically active miR-21, determined using a fusion mRNA reporter system carrying miR-21 target sequences in its 3'UTR. Upregulation of miR-21 expression increased aldosterone secretion but not cortisol secretion. Elevation of miR-21 levels promoted H295R cell proliferation. Deregulation in miR-21 expression may be involved in dysregulation of angiotensin II signaling and abnormal aldosterone secretion by adrenal glands in humans. This study raises the possibility that increased miR-21 expression levels outside of the heart may contribute to adverse cardiac remodeling through dysregulation of angiotensin II–mediated signaling and enhanced aldosterone secretion [Divakaran & Mann, 2008].

miR-155

The adverse effects of angiotensin II on heart failure are primarily mediated through the angiotensin II type 1 receptor (AT1R). A study reported by Martin *et al* [2006] establish the ability of miR-155 to suppress the levels of AT1R. Functional studies demonstrated that transfection of miR-155 into human primary lung fibroblasts (hPFBs) reduced the endogenous expression of the AT1R compared with non-transfected cells. Luciferase reporter assays confirmed that miR-155 could directly interact with the 3'UTR of the AT1R mRNA. Additionally, miR-155 transfected cells showed a significant reduction in angiotensin II-induced extracellular signal-related kinase 1/2 (ERK1/2) activation. Furthermore, when hPFBs were transfected with an antisense to miR-155, endogenous AT1R expression and angiotensin II-induced ERK1/2 activation were significantly increased. Finally, transforming growth factor-beta1 treatment of hPFBs resulted in the decreased expression of miR-155 and the increased expression of the AT1R. The results suggest that miR-155 can translationally repress the expression of AT1R protein *in vivo*. Importantly, the translational repression mediated by miR-155 can be regulated by physiological stimuli.

In the subsequent study conducted by the same group [Martin *et al.*, 2007], the authors further demonstrated the ability of miR-155 to target a silent polymorphism (+1166 A/C) in the human AT1R gene that has been associated with cardiovascular disease, possibly as a result of enhanced AT1R activity. Because this polymorphism occurs in the 3'UTR of the human AT1R gene, the biological importance of this mutation has always been questionable. Computer alignment revealed that the +1166 A/C polymorphism occurred in a *cis*-regulatory site, which is recognized by miR-155. When the +1166 C-allele is present, base-pairing complementarity is interrupted, and the ability of miR-155 to interact with the *cis*-regulatory site is decreased. As a result, miR-155 no longer attenuates translation as efficiently as demonstrated by luciferase reporter and angiotensin II radioreceptor binding assays. In situ hybridization experiments demonstrated that mature miR-155 is abundantly expressed in the same cell types as the AT1R (e.g. endothelial and vascular smooth muscle). Finally, when human primary vascular smooth muscle cells were transfected with a miR-155 AMO, endogenous human AT1R expression and Ang II-induced ERK1/2 activation were significantly increased. Clearly, the AT1R and miR-155 are co-expressed and miR-155 translationally represses the expression of AT1R *in vivo*. This study provides a feasible biochemical mechanism by which the +1166 A/C polymorphism can lead to increased AT1R densities and possibly cardiovascular disease.

REFERENCES

Cheng Y, Ji R, Yue J, Yang J, Liu X, Chen H, Dean DB, Zhang C. (2007) MicroRNAs are aberrantly expressed in hypertrophic heart. Do they play a role in cardiac hypertrophy? Am J Pathol 170:1831–1840.

Divakaran V, Mann DL. (2008) The emerging role of microRNAs in cardiac remodeling and heart failure. Circ Res 103:1072–1083.

Mann DL, Bristow MR. (2005) Mechanisms and models in heart failure: the biomechanical model and beyond. Circulation 111:2837–2849.

Martin MM, Buckenberger JA, Jiang J, Malana GE, Nuovo GJ, Chotani M, Feldman DS, Schmittgen TD, Elton TS. (2007) The human angiotensin II type 1 receptor +1166 A/C polymorphism attenuates microrna-155 binding. J Biol Chem 282:24262–24269.

Martin MM, Lee EJ, Buckenberger JA, Schmittgen TD, Elton TS. (2006) MicroRNA-155 regulates human angiotensin II type 1 receptor expression in fibroblasts. *J Biol Chem* 28:18277–18284.

Romero DG, Plonczynski MW, Carvajal CA, Gomez-Sanchez EP, Gomez-Sanchez CE. (2008) Microribonucleic acid-21 increases aldosterone secretion and proliferation in H295R human adrenocortical cells. Endocrinology 149:2477–2483.

Sayed D, Hong C, Chen IY, Lypowy J, Abdellatif M (2007) MicroRNAs play an essential role in the development of cardiac hypertrophy. Circ Res 100:416–424.

Tatsuguchi M, Seok HY, Callis TE, Thomson JM, Chen JF, Newman M, Rojas M, Hammond SM, Wang DZ. (2007) Expression of microRNAs is dynamically regulated during cardiomyocyte hypertrophy. J Mol Cell Cardiol 42:1137–1141.

CHAPTER 16

miRNAs in Cardiac Metabolism

Abstract: This chapter aims to provide an implicit introduction to the role of miRNAs in regulating cardiac metabolism. The homeostasis of glucose, lipid, protein, and energy, which is critical for normal cardiovascular function, is maintained by cellular metabolism. Metabolic perturbation occurs in various types of cardiac disease, including myocardial ischemia, cardiac hypertrophy, heart failure, diabetic cardiomyopathy, atherosclerosis, etc. The depletion of high-energy-phosphate metabolites may contribute to heart failure, and a decreased PCr/ATP ratio has been found in cardiac muscle of heart failure patients and animal models of heart failure. A major determinant of glycolytic flux is glucose transport; glucose enters cardiac cells via the facilitative glucose transporters GLUT1 and GLUT4. Several miRNAs have been demonstrated to produce regulatory effects on GLUT4, cellular ATP level, and the pleiotropic factor IGF-1. A succinct summary on these studies is given in this chapter.

INTRODUCTION

The homeostasis of glucose, lipid, protein, and energy, which is critical for normal cardiovascular function, is maintained by cellular metabolism. Metabolic perturbation occurs in various types of cardiac disease, including myocardial ischemia, cardiac hypertrophy, heart failure, diabetic cardiomyopathy, atherosclerosis, etc. Moreover, metabolic remodeling that involves adverse alterations of every aspects of metabolism is a characteristic of progression of the pathologic processes. For example, myocardial ischemia leads to a cascade of metabolic events which are interrelated and are caused by hypoxia, acidosis, oxidative stress, calcium overload, decreases in survival signaling molecules and increases in death signaling mediators, etc. The ischemic injuries are typically (1) contractile dysfunction, (2) electrical disturbance, and (3) cell loss.

miRNAs AND GLUCOSE METABOLISM

miR-133 & KLF15/GLUT4

The depletion of high-energy-phosphate metabolites may contribute to heart failure, and a decreased PCr/ATP ratio has been found in cardiac muscle of heart failure patients [Neubauer *et al.*, 1992] and animal models of heart failure [Horn *et al.*, 2001]. Myocardial glycolytic ATP production is important for preserving cardiac viability during ischemia [Cross *et al.*, 1996]. A major determinant of glycolytic flux is glucose transport; glucose enters cardiac cells via the facilitative glucose transporters GLUT1 and GLUT4, the latter of which is the major mechanism by which glucose uptake into the cardiomyocytes [Tian *et al.*, 2001]. GLUT4 level is decreased in failing human adult hearts [Stanley *et al.*, 2005]. Moreover, type 2 diabetes mellitus (T2DM) is a metabolic disorder characterized by insulin resistance in the whole body and myocardial level [Paternostro *et al.*, 1996; Dutka *et al.*, 2006].1,2 Patients with T2DM are at increased risk of developing cardiovascular disease and have a significantly increased risk of left ventricular dysfunction (LVD) [Nathan *et al.*, 1997; Grundy *et al.*, 1999], which is independently associated with myocardial insulin resistance.

Horie *et al* [2009] found that forced expression of miR-133 decreased GLUT4 expression and reduced insulin-mediated glucose uptake in cardiomyocytes. The expression of miR-133a and miR-133b also reduced the protein levels of KLF15, whereas they did not change the KLF15 mRNA level, which reduced the level of the downstream target GLUT4. Cardiac myocytes infected with lenti-decoy, in which the 3'UTR with tandem antisense sequence to miR-133 was linked to the luciferase reporter gene, had decreased miR-133 levels and increased levels of GLUT4. The expression levels of KLF15 and GLUT4 were decreased at the left ventricular hypertrophy and congestive heart failure stage in a rat model. The expression levels of GLUT4 and miR-133 were also investigated in Dahl salt-sensitive rats [Inoko *et al.*, 1994]. In this rat model under a high-salt diet, systemic hypertension caused compensated concentric left ventricular hypertrophy (LVH) at the age of 11 weeks, followed by marked LV dilatation and global hypokinesis at 17 weeks (CHF stage). The expression levels of KLF15 and GLUT4 in the heart were reduced at the LVH stage and further reduced at the CHF stage. The results indicated that miR-133 regulates the expression of GLUT4 by targeting KLF15 and is involved in metabolic control in cardiomyocytes.

Together with our previous findings and findings from other laboratories that miR-133 is upregulated in a rabbit model of diabetes but downregulated in cardiac hypertrophy/heart failure [Xiao *et al.*, 2007; Luo *et al.*, 2008;], it appears that miR-133 may contribute to the dysfunction of GLUT4 in diabetic cardiomyopathy, in addition to its adaptive role to minimize GLUT4 downregulation in heart failure.

miR-223 & GLUT4

More direct evidence for the role of miRNAs in regulating GLUT4 came from a study conducted with left ventricular biopsies collected from patients with or without type 2 diabetes and from patients with left ventricular dysfunction [Lu *et al.*, 2010]. But in this study, miRNA was found to produce post-transcriptional activation of GLUT4 expression. Quantitative miRNA expression analyses of 155 miRNAs revealed that miR-223 was consistently upregulated in the insulin-resistant heart. The effects of miR-223 on glucose metabolism in neonatal rat cardiomyocytes where adenoviral-mediated overexpression of miR-223 increased glucose uptake were assessed. They showed that independent of PI3K or AMPK activity, miR-223 increases cellular glucose uptake via Glut4. This increase in glucose uptake was inhibited by using siRNA against Glut4. It is notable that following siRNA against Glut4, glucose uptake in miR-223 overexpressing cells fell below that of control cells. Using in silico miRNA target prediction programs, the authors prioritized candidate miR-223 target genes, but observed no effect of miR-223 on myocyte enhancer factor 2c or insulin-like growth factor 1 receptor, and an unexpected miR-223-induced increase in nuclear factor IA. They next examined the effects of miR-223 on insulin signalling and glucose transport proteins. Neither phosphoinositide 3-kinase (PI3K) signalling nor AMP kinase activity was affected by miR-223 overexpression, whereas glucose transporter 4 (Glut4) protein expression was increased. Thus, miR-223 overexpression-induced Glut4 protein expression in cardiomyocytes was necessary and sufficient for increased glucose uptake as demonstrated by siRNA knockdown of Glut4. Loss-of-function studies *in vivo*, using a synthetic miR-223 inhibitor, confirmed the effect of miR-223 on Glut4.

miR-15b AND CELLULAR ATP LEVEL

Mitochondria constitute ~40% of the total cardiomyocyte volume in the heart [Barth *et al.*, 1992], a feul plant which generates cellular ATP for use in cellular metabolism. To identify miRNAs that can affect cellular ATP levels and mitochondria, Nishi and colleagues [Nishi *et al.*, 2010] conducted an interesting study in which a series of highly expressed miRNAs in the heart were individually overexpressed in neonatal rat cardiomyocytes using a lentiviral vector: including let-7d, miR-1, miR-15b, miR-16, miR-21, miR-22, miR-23a, miR-27a, miR-29a, miR-125a, miR-126–3p, miR-133a, miR-133b, miR-143, miR-144, miR-146, miR-149, miR-155, miR-181, miR-195, miR-199a, miR-214, and miR-424. They found that miR-15b decreased cellular ATP levels without affecting cell viability. miR-16, -195, and -424, which have the same seed sequence as miR-15b, also decreased cellular ATPlevels. Through a bioinformatics approach, ADP-ribosylation factor-like 2 (Arl2), which is known to localize to adenine nucleotide transporter 1, the exchanger of ADP/ATP in mitochondria [Chen *et al.*, 1999], was identified as a potential target of miR-15b. Overexpression of miR-15b, miR-16, miR-195, and miR-424 suppressed the activity of a luciferase reporter construct fused with the 3'-untranslated region of Arl2. In addition, miR-15b overexpression decreased Arl2 mRNA and protein expression levels. The effects of Arl2 siRNA on cellular ATP levels were the same as those of miR-15b, and the expression of Arl2 could restore ATP levels reduced by miR-15b. A loss-of-function study of miR-15b resulted in increased Arl2 protein and cellular ATP levels. Electron microscopic analysis revealed that mitochondria became degenerated in cardiomyocytes that had been transduced with miR-15b and Arl2 siRNA. The results suggest that miR-15b may decrease mitochondrial integrity by targeting Arl2 in the heart.

miR-1 & IGF-1

In myocardial biology, IGF-1, a pleiotropic factor, and its signal transduction cascade are involved in the control of virtually every critical biological process, including development, cardiomyocyte size and survival, action potential, antiapoptotic [Santini *et al.*, 2007; Muta & Krantz, 1993; Matthews & Feldman, 1996; Jung *et al.*, 1996], hypertrophic, and regenerative effects on cardiac muscle [Torella *et al.*, 2004; Lembo *et al.*, 1996], and excitation-contraction coupling [Latronico *et al.*, 2004; Santini *et al.*, 2007]. Binding of IGF-1 to its receptor activates its intrinsic receptor tyrosine kinase, which phosphorylates several intracellular substrates such as the insulin receptor substrate-1 and Shc, leading to the activation of signaling pathways, including phosphatidylinositol 3-kinase/AKT pathways [Brunet *et al.*, 1999].

Condorelli's group [Elia *et al.*, 2009] recently identified the role of miR-1 in regulating expression of IGF-1. Utilizing bioinformatics tools, biochemical assays, and *in vivo* models, they demonstrated that (1) insulin-like growth factor-1 (IGF-1) and IGF-1 receptor are targets of miR-1; (2) miR-1 and IGF-1 protein levels are correlated inversely in models of cardiac hypertrophy and failure as well as in the C2C12 skeletal muscle cell model of differentiation; (3) the activation state of the IGF-1 signal transduction cascade reciprocally regulates miR-1 expression through the Foxo3a transcription factor; and (4) miR-1 expression correlates inversely with cardiac mass and thickness in myocardial biopsies of acromegalic patients, in which IGF-1 is overproduced after aberrant synthesis of growth hormone.

REFERENCES

Barth E, Stämmler G, Speiser B, Schaper J. (1992) Ultrastructural quantitation of mitochondria and myofilaments in cardiac muscle from 10 different animal species including man. J Mol Cell Cardiol 24:669-681.

Brunet A, Bonni A, Zigmond MJ, Lin MZ, Juo P, Hu LS, Anderson MJ, Arden KC, Blenis J, Greenberg ME. (1999) Akt promotes cell survival by phosphorylating and inhibiting a Forkhead transcription factor. Cell 96:857–868.

Carè A, Catalucci D, Felicetti F, Bonci D, Addario A, Gallo P, Bang ML, Segnalini P, Gu Y, Dalton ND, Elia L, Latronico MV, Høydal M, Autore C, Russo MA, Dorn GW, Ellingsen O, Ruiz-Lozano P, Peterson KL, Croce CM, Peschle C, Condorelli G. (2007) MicroRNA-133 controls cardiac hypertrophy. Nat Med 13:613–618.

Chen X, Van Valkenburgh C, Fang H, Green N. (1999) Signal peptides having standard and nonstandard cleavage sites can be processed by Imp1p of the mitochondrial inner membrane protease. J Biol Chem 274:37750–37754.

Cross HR, Opie LH, Radda GK, Clarke K. (1996) Is a high glycogen content beneficial or detrimental to the ischemic rat heart? A controversy resolved. Circ Res 78:482–491.

Dutka DP, Pitt M, Pagano D, Mongillo M, Gathercole D, Bonser RS, Camici PG. (2006) Myocardial glucose transport and utilization in patients with type 2 diabetes mellitus, left ventricular dysfunction, and coronary artery disease. J Am Coll Cardiol 48:2225–2231.

Elia L, Contu R, Quintavalle M, Varrone F, Chimenti C, Russo MA, Cimino V, De Marinis L, Frustaci A, Catalucci D, Condorelli G. (2009) Reciprocal regulation of microRNA-1 and insulin-like growth factor-1 signal transduction cascade in cardiac and skeletal muscle in physiological and pathological conditions. Circulation 120:2377–2385.

Grundy SM, Benjamin IJ, Burke GL, Chait A, Eckel RH, Howard BV, Mitch W, Smith SC Jr, Sowers JR. (1999) Diabetes and cardiovascular disease: a statement for healthcare professionals from the American Heart Association. Circulation 100:1134–1146.

Horie T, Ono K, Nishi H, Iwanaga Y, Nagao K, Kinoshita M, Kuwabara Y, Takanabe R, Hasegawa K, Kita T, Kimura T. (2009) MicroRNA-133 regulates the expression of GLUT4 by targeting KLF15 and is involved in metabolic control in cardiac myocytes. Biochem Biophys Res Commun 389:315–320.

Horn M, Remkes H, Strömer H, Dienesch C, Neubauer S. (2001) Chronic phosphocreatine depletion by the creatine analogue beta-guanidinopropionate is associated with increased mortality and loss of ATP in rats after myocardial infarction. Circulation 104:1844–1849.

Inoko M, Kihara Y, Morii I, Fujiwara H, Sasayama S. (1994) Transition from compensatory hypertrophy to dilated, failing left ventricles in Dahl salt-sensitive rats. Am J Physiol 267:H2471–H2482.

Jung Y, Miura M, Yuan J. (1996) Suppression of interleukin-1 beta-converting enzyme-mediated cell death by insulin-like growth factor. J Biol Chem 271:5112–5117.

Latronico MV, Costinean S, Lavitrano ML, Peschle C, Condorelli G. (2004) Regulation of cell size and contractile function by AKT in cardiomyocytes. Ann N Y Acad Sci 1015:250–260.

Lembo G, Rockman HA, Hunter JJ, Steinmetz H, Koch WJ, Ma L, Prinz MP, Ross J Jr, Chien KR, Powell-Braxton L. (1996) Elevated blood pressure and enhanced myocardial contractility in mice with severe IGF-1 deficiency. J Clin Invest 98:2648–2655.

Lu H, Buchan RJ, Cook SA. (2010) MicroRNA-223 regulates Glut4 expression and cardiomyocyte glucose metabolism. Cardiovasc Res 86:410–420.

Luo X, Lin H, Xiao J, Zhang Y, Lu Y, Yang B, Wang Z. (2008) Downregulation of *miRNA-1/miRNA-133* contributes to re-expression of pacemaker channel genes *HCN2* and *HCN4* in hypertrophic heart. J Biol Chem 283:20045–20052.

Matthews CC, Feldman EL. (1996) Insulin-like growth factor I rescues SH-SY5Y human neuroblastoma cells from hyperosmotic induced programmed cell death. J Cell Physiol 166:323–331.

Muta K, Krantz SB. (1993) Apoptosis of human erythroid colony-forming cells is decreased by stem cell factor and insulin-like growth factor I as well as erythropoietin. J Cell Physiol 156:264–271.

Nathan DM, Meigs J, Singer DE. (1997) The epidemiology of cardiovascular disease in type 2 diabetes mellitus: how sweet it is . . . or is it? Lancet 350(Suppl 1):SI4–SI9.

Neubauer S, Krahe T, Schindler R, Horn M, Hillenbrand H, Entzeroth C, Mader H, Kromer EP, Riegger GA, Lackner K. (1992) 31P magnetic resonance spectroscopy in dilated cardiomyopathy and coronary artery disease. Altered cardiac high-energy phosphate metabolism in heart failure. Circulation 86:1810–1818.

Nishi H, Ono K, Iwanaga Y, Horie T, Nagao K, Takemura G, Kinoshita M, Kuwabara Y, Mori RT, Hasegawa K, Kita T, Kimura T. (2010) MicroRNA-15b modulates cellular ATP levels and degenerates mitochondria via Arl2 in neonatal rat cardiac myocytes. J Biol Chem 285:4920-4930.

Paternostro G, Camici PG, Lammerstma AA, Marinho N, Baliga RR, Kooner JS, Radda GK, Ferrannini E. (1996) Cardiac and skeletal muscle insulin resistance in patients with coronary heart disease. A study with positron emission tomography. J Clin Invest 98:2094–2099.

Santini MP, Tsao L, Monassier L, Theodoropoulos C, Carter J, Lara-Pezzi E, Slonimsky E, Salimova E, Delafontaine P, Song YH, Bergmann M, Freund C, Suzuki K, Rosenthal N. (2007) Enhancing repair of the mammalian heart. Circ Res 100:1732–1740.

Stanley WC, Recchia FA, Lopaschuk GD. (2005) Myocardial substrate metabolism in the normal and failing heart. Physiol Rev 85:1093–1129.

Tian R, Abel ED. (2001) Responses of GLUT4-deficient hearts to ischemia underscore the importance of glycolysis. Circulation 103:2961–2966.

Torella D, Rota M, Nurzynska D, Musso E, Monsen A, Shiraishi I, Zias E, Walsh K, Rosenzweig A, Sussman MA, Urbanek K, Nadal-Ginard B, Kajstura J, Anversa P, Leri A. (2004) Cardiac stem cell and myocyte aging, heart failure, and insulin-like growth factor-1 overexpression. Circ Res 94:514–524.

Xiao J, Luo X, Lin H, Zhang Y, Lu Y, Wang N, Zhang YQ, Yang B, Wang Z. (2007) MicroRNA *miR-133* represses HERG K$^+$ channel expression contributing to QT prolongation in diabetic hearts. J Biol Chem 282:12363–12367.

CHAPTER 17

Circulating miRNAs as Biomarkers for Cardiac Disease

Abstract: This chapter aims to discuss recent advances of circulating miRNAs as new and promising biomarkers for cardiac disease. The elucidation of miRomes between diseased and normal cardiovascular tissues or between different cardiovascular disease types, stages and grades, gives the chance to identify the miRNAs most probably involved in cardiovascular disease and to establish new diagnostic and prognostic markers. Recent findings suggest that circulating miRNAs may be plasma biomarkers for the diagnosis of lung, colorectal, and prostate cancers. These findings have been also tested for cardiovascular disease. miRNAs are present in human plasma in a remarkably stable form that is protected from endogenous RNase activity. The levels of miRNAs in serum are reproducible and consistent among individuals of the same species. In particular, blood miR-1, miR-133, miR-208a and miR-499 have been suggested as biomarkers of acute myocardial infarction; miR-208, miR-423-5p and some other miRNAs in the circulation are correlated with heart failure; and miR-122, miR-124 and miR-133 may be used to predict cerebral artery occlusion stroke.

INTRODUCTION

Besides being recognized as key molecules in intracellular regulatory networks for gene expression, the spectra and levels of some miRNAs are emerging as biomarkers for various pathological conditions [Waldman & Terzic 2008; Dillhoff *et al* 2008]. Recent findings suggest that circulating miRNAs may be plasma biomarkers for the diagnosis of lung [Chen *et al* 2008], colorectal [Chen *et al* 2008], and prostate cancers [Mitchell *et al* 2008]. Indeed, miRNAs are present in the serum and plasma of humans and other animals such as mice, rats, bovine fetuses, calves, and horses [Ai *et al* 2009; Chen *et al* 2008; Hunter *et al* 2008; Mitchell *et al* 2008; Chim *et al* 2008; Gilad *et al* 2008; Lawrie *et al* 2008; Resnick *et al* 2008; Taylor *et al* 2008; Wang *et al* 2009].

The elucidation of miRomes between diseased and normal cardiovascular tissues or between different cardiovascular disease types, stages and grades, gives the chance to identify the miRNAs most probably involved in cardiovascular disease and to establish new diagnostic and prognostic markers. An early and correct diagnosis may warrant immediate initiation of reperfusion therapy to potentially reduce the mortality rate. An ideal biomarker should be abundantly and preferentially expressed in the tissue of interest, and be typically present at low concentrations in the blood and other body fluids. Upon tissue injury, such biomarkers are released into the systemic circulation, where they can be detected in a blood-based assay. Biomarkers, used to establish a diagnosis in patients with heart disease, have emerged largely from targeted analyses of known myocardial proteins and become more and more important for diagnosis of AMI [Thygesen *et al*., 2007; Jaffe *et al*., 2006]. Current biomarkers such as creatine kinase–MB isoenzymes, cardiac myoglobin, and troponins have been widely applied in clinical diagnosis [de Winter *et al*., 1995]. Among these, cardiac troponins are currently considered as the 'gold standard' for acute myocardial infarction diagnosis [Jaffe *et al*., 2000]. Clinical management of heart disease is facilitated by circulating biomarkers like brain natriuretic peptide (BNP) [Januzzi *et al*., 2006; van Kimmenade *et al*., 2008].

Challenges for developing protein-based biomarkers from body fluids, such as plasma, serum, and urine, include the complexity of protein composition, the assorted posttranslational modifications, the low abundance of proteins of interest, the difficulty of developing suitable high-affinity capture agents, the complexities of proteolysis and protein denaturation, and potentially complex assay methods. All of these make the discovery and development of protein-based biomarkers with proper specificity, sensitivity, and predictive value an expensive, time consuming, and difficult task. The biological function of circulating miRNA is largely unknown; however, unlike proteins, there are far fewer known miRNA species, so obtaining a complete profile is relatively easy. Currently, there are 866 known miRNAs for human and 627 for mouse (based on the latest miRBase release 12.0 http://microrna.sanger.ac.uk/sequences/index.shtml) compared with perhaps a million or more serum proteins, including various processing variants and posttranslationally modified proteins. mRNA must be translated into protein to have a biological effect whereas miRNAs are themselves the active moiety, often influencing the expression of multiple other genes, and thus likely reflect altered physiology more directly. In addition, miRNAs do not have known post-processing modifications, and with their size, their chemical composition is much less complex than most other biological molecules. Detecting specific miRNA species, although somewhat challenging, is inherently a much easier task than detecting proteins. A synthetic complementary oligonucleotide should deliver sufficient specificity in most cases, and a standard PCR assay can be used to increase the detection sensitivity. It has also been demonstrated that the circulating miRNAs are stable and can be reliably extracted and assayed in either serum or plasma [Mitchell *et al* 2008]. Both real-time RT-PCR and microarray methods have been applied to detect blood miRNAs.

miRNAs are present in human plasma in a remarkably stable form that is protected from endogenous RNase activity. The levels of miRNAs in serum are reproducible and consistent among individuals of the same species. Strikingly, Gilad *et al* [Gilad *et al* 2008] confirmed that miRNAs are also detectable in other body fluids, such as urine, saliva, amniotic fluid and pleural fluid. Of note, serum and urine display different miRNA abundance profiles as might be expected for two dissimilar biological fluids, further supporting the hypothesis that bodily fluid microRNA profiles reflect physiology.

(1) Even with much less degree of freedom than mRNAs, miRNA expression profiles reflect the developmental lineage and differentiation state of cells and successfully classified poorly differentiated tumors that could not have definitive diagnosis by histopathology, while the classification based upon the mRNA profiles was highly inaccurate.

(2) A very promising diagnostic strategy could arise from miRNAs if they are found in serum and can be detected by RT-PCR. Recent studies clearly indicate that miRNAs have unusually high stability in formalin-fixed, paraffin-embedded tissues, can remain intact in plasma and serum as well. Indeed, while expression patterns of miRNAs in tissue specimens have been well regarded as better biomarkers of human cancer, being characteristic of tumor type, tumor grade and developmental origin, most recently circulating miRNAs have also been revealed to be the stable blood-based markers for cancer detection. This is presumably because miRNAs are shorter than mRNAs, and therefore more resistant to ribonuclease degradation.

(3) Different diseases have distinct miRNA transcriptomes. The cancer cluster of miRNAs is clearly separated from the cardiovascular disease cluster of miRNAs.

(4) Studies indicate that all cancers are connected together by miRNA profiles, suggesting that various cancers may share similar associations at the miRNA level, in which some strong onco-miRNAs or miRNA tumor suppressors may play key roles. In a remarkable study, Lu *et al* [Lu *et al* 2005] showed that expression data for 217 miRNAs only, performed better at identifying cancer types than analysis of 16,000 mRNAs. They concluded that miRNAs might help detecting cancer better than other strategies presently available because miRNAs are only several hundreds, compared to tens of thousands of mRNAs and proteins. This can also partly be attributed to the fact that miRNA expression tends to be very strictly defined in time and space [Xi *et al* 2006].

(5) This result revealed a potential correlation between miRNA tissue specificity and disease, which may be of value in predicting specific diseaserelated miRNAs by combining the miRNA tissue specificity values. Thus, if a disease occurs specifically in a given tissue, the miRNAs specifically expressed in that tissue will have a great potential to be related to that disease.

(6) Disease-associated miRNAs show various dysfunctions, such as mutation, up-regulation, deleted, and down-regulation.

(7) Finally, miRNA analysis requires no expensive and time-consuming detection strategies using antibodies or mass spectrometry.

CIRCULATING miRNAs AND ACUTE MYOCARDIAL INFARCTION

Wang *et al* [2010] reported an interest study aiming to identify the miRNA biomarkers for acute myocardial infarction (AMI). They first obtained the miRNA transcroptome in the plasma of healthy human subjects: some 170 miRNAs were detected by microarray. They noticed that miR-133a had a low level in the plasma, and miR-1, miR-499 and miR-208a were undetectable. Real-time RT-PCR analysis revealed that miR-451 and miR-16 were at high abundance, while miR-133a was at a low level. In addition, miR-1 and miR-499 could be detected with a marginal level of expression. miR-208a could be detected neither by microarray nor by real-time PCR.

This study assessed circulating levels of two cardiac specific miRNA miR-208 and miR-499, and two muscle-specific miRNAs miR-1 and miR-133 in the plasma from 33 consecutive AMI and 33 non-AMI patients. Quantitative RT-PCR analysis that all four miRNA levels were substantially higher than those from healthy people, patients with non-AMI coronary heart disease, or patients with other cardiovascular diseases. miR-1, miR-133a, and miR-499 were detected with higher levels in plasma from the AMI group compared with those from the healthy group. Most notably, miR-208a remained undetectable in non-AMI patients, but was readily detected in 90.9% AMI patients and in 100% AMI patients within 4 h of the onset of symptoms. By receiver operating characteristic curve analysis, among the four miRNAs investigated, miR-208a revealed the higher sensitivity and specificity for diagnosing AMI. This study therefore provides the first clinical evidence of circulating miR-208a as a biomarker of cardiac damage.

To see if a similar profile of miRNAs could be presented in experimental AMI, the authors continued to compare the selected plasma miRNA levels in a rat model of AMI created by coronary artery ligation. Blood samples were collected from the rats at various time points after coronary artery ligation. Real-time RT-PCR analysis showed that the levels of miR-1, miR-133a, and miR-499 were increased at 1–3 h, peaked at 3–12 h and decreased at the 12–24 h time point. The level of miR-208a was undetectable at 0 h, but significantly increased to a detectable level at as early as 1 h, peaked at 3 h, then began to decrease at 6–12 h and reduced to the undetectable level at 24 h after coronary artery occlusion. The changes were therefore quite comparable to human subjects.

A more recent study recruited 159 patients with or without AMI for quantification of miR-1 level in plasma using real-time RT-PCR method [Ai *et al.*, 2010]. The authors found that miR-1 level was significantly higher in plasma from AMI patients compared with non-AMI subjects and the level was dropped to normal on discharge following medication. Increased circulating miR-1 was not associated with age, gender, blood pressure, diabetes mellitus or the established biomarkers for AMI. However, miR-1 level was well correlated with QRS by both univariable linear and logistics regression analyses. The area under ROC curve (AUC) was 0.7740 for separation between non-AMI and AMI patients and 0.8522 for separation AMI patients under hospitalization and discharge. Receiver-operator characteristic curve (ROC) analysis indicate a strong diagnostic ability of circulating miR-1 may be a novel, independent biomarker for diagnosis of AMI. Consistently, we have previously documented upregulation of miR-1 in ischemic heart of both patients and rats [Yang *et al.*, 2007]. This cardiac upregulation suggests that increased circulating miR-1 in myocardial infarction is derived from the myocardium resulted from increased myocardial production and subsequent release from the heart.

CIRCULATING miRNAs AND HEART FAILURE

A mechanistic role for altered mRNA expression levels in heart disease has been recognized for many years [Chien *et al.*, 1991]. A few hallmark genes are regulated in virtually every clinical and experimental model of cardiac hypertrophy and/or heart failure. The most sensitive transcriptional marker for heart failure is increased cardiomyocyte expression of mRNAs for the atrial and brain natriuretic peptides, ANF and BNP. On the other hand, cardiomyocyte hypertrophy is indicated by a redistribution of myosin heavy chain isoform mRNA from alpha (α-MHC) to beta (β-MHC). Transcriptional upregulation of natriuretic peptides and β-MHC is observed in cultured cardiac myocytes induced to undergo hypertrophy, in genetic rodent models of cardiac hypertrophy and cardiomyopathy, in rabbit, dog, and porcine experimental models of surgically induced cardiac disorders, and in the analogous human diseases. Thus, a conserved transcriptional signature for heart disease appears to be a nearly universal response, and in many instances the individual regulated transcripts have been mechanistically connected to specific pathological features of hypertrophied and failing myocardium [Dorn et a., 2003]. However, because end-stage cardiomyopathy combines features of heart failure with cardiomyocyte hypertrophy, ANP/BNP and β-MHC are typically increased together with many other members of the so-called 'fetal gene program' in fully developed adult cardiomyopathies. Combinatorial regulation of many mRNAs in cardiac disease diminishes the specificity of the response for a particular condition. For example upregulation of ANP along with β-MHC in a cardiomyopathic heart does not provide data as to whether cardiomyocyte hypertrophy led to heart failure, as in severe hypertension, or is part of a compensatory response to primary myocardial damage, as after myocardial infarction. Furthermore, since heart failure is the common terminal condition that results from irreparable myocardial damage of any cause, the transcriptional profile of late heart failure provides little insight into specific aetiology (ischaemic, viral, alcoholic) or information about likely prognosis [Margulies *et al.*, 2009]. For these reasons, mRNA profiling has not transitioned from the research laboratory to routine clinical practice [Dorn & Matkovich, 2008]. There is, however, tremendous interest in determining whether dynamic regulation of cardiac-expressed miRNAs could prove more useful than mRNA profiling as a molecular signature for specific cardiac syndromes. With the development of faster and cheaper highthroughput technologies, this could be a particularly exciting prospect for diagnosis and prognostication in cardiology.

In an animal study, miR-208 was found to be a reasonable biomarker for myocardial injury induced by isoproterenol in rat [Ji *et al.*, 2009]. Plasma concentration of miR-208 was found increased significantly after isoproterenol-induced myocardial injury and showed a similar time course to the concentration of cTnI, a classic biomarker of myocardial injury.

In a more recent study in human subject, a miRNA array was performed in 12 healthy controls and 12 patients with heart failure [Tijsen *et al.*, 2010]. From this array, the authors selected 16 miRNAs for a second clinical study in 39 healthy controls and in 50 cases with reports of dyspnea, of whom 30 were diagnosed with heart failure and 20 were diagnosed with dyspnea attributable to non–heart failure-related causes. The results revealed that miR-423-5p was specifically enriched in blood of heart failure cases and receiver-operator-characteristics (ROC) curve analysis showed miR-423-5p to be a diagnostic predictor of heart failure, with an area under the curve of 0.91. Consistely, miR-423-5p has been reported in array studies to be upregulated in human failing myocardium [Thum *et al.*, 2007]. Five other miRNAs were elevated in heart failure cases but also slightly increased in non-heart failure dyspnea cases. In addition, they also identify another 6 miRNAs that are elevated in patients with heart failure, including miR-18b*, miR-129–5p, miR-1254, miR-675, HS_202.1 and miR-622. Within the dyspnea population, miR-18b* is also slightly elevated in non-heart failure cases, whereas miR-675 is even upregulated in non-heart failure cases to the same level as heart failure cases. Finally, some candidate miRNAs (miR-423-5p and miR-675, but not miR-18b*) were higher in atherosclerotic forms of heart failure as compared to nonatherosclerotic forms of heart failure compare heart failure cases not only to controls, but also to dyspneic patients who were free of clinical heart failure. This enables distinguishing between miRNAs that are upregulated in clinical heart failure and those that are upregulated more in general with dyspnea. An example of this is miR-675. This miRNA seemed an attractive candidate when only comparing heart failure cases to fully healthy controls, but appears to be generally upregulated in dyspnea, and not specific for heart failure. Thus, these 6 circulating miRNAs, in particular miR-423-5p, provide attractive candidates as putative biomarkers for heart failure. It should be noted, however, that this study is limited by the relatively small number of patients.

miRNAs AND CEREBRAL ARTERY OCCLUSION STROKE

Biomarkers of brain injury that are most commonly cited in the literature include S-100B, neuron-specific enolase, glial fibrillary acidic protein, myelin basic protein, and fatty acid– binding protein [Rothermundt *et al.*, 2003; Vaage & Anderson, 2001; Persson *et al.*, 1987; Anand & Stead, 2005]. Most of these biomarkers, however, are considered to lack specificity for brain injury. Laterza *et al* [2009] used a surgical model of stroke that involved occluding the middle cerebral artery of rat. Transient (60–90 min) and permanent occlusions were performed to obtain different levels of injury. They then assessed plasma samples for the presence of the brain specific miR-124 [Liang *et al.*, 2007] according to the time after occlusion, as well as the liver- and muscle-specific miRNAs miR-122 and miR-133a. Concentrations of miR-124 were increased in plasma beginning at 8 h, but only in the transient occlusion/ischemia samples, whereas baseline values were observed for both miR-122 and miR-133a in the samples. Interestingly, this early increase in miR-124 was not seen in rats that had undergone permanent occlusion. Larger increases in miR-124 concentrations (up to 150-fold compared with sham-operated rats) were seen 24 h after stroke for both the transient and permanent occlusions.

REFERENCES

Ai J, Zhang R, Li Y, Pu J, Lu Y, Jiao J, Li K, Yu B, Li Z, Wang R, Wang L, Li Q, Wang N, Shan H, Li Z, Yang B. (2010) Circulating microRNA-1 as a potential novel biomarker for acute myocardial infarction. Biochem Biophys Res Commun 391:73–77.

Anand N, Stead LG. (2005) Neuron-specific enolase as a marker for acute ischemic stroke: a systematic review. Cerebrovasc Dis 20:213–219.

Chen X, Ba Y, Ma L, Cai X, Yin Y, Wang K, Guo J, Zhang Y, Chen J, Guo X, Li Q, Li X, Wang W, Zhang Y, Wang J, Jiang X, Xiang Y, Xu C, Zheng P, Zhang J, Li R, Zhang H, Shang X, Gong T, Ning G, Wang J, Zen K, Zhang J, Zhang CY. (2008) Characterization of microRNAs in serum: a novel class of biomarkers for diagnosis of cancer and other diseases. Cell Res 18:997–1006.

Chien KR, Knowlton KU, Zhu H, Chien S. (1991) Regulation of cardiac gene expression during myocardial growth and hypertrophy: molecular studies of an adaptive physiologic response. FASEB J 5:3037–3046.

Chim SS, Shing TK, Hung EC, Leung TY, Lau TK, Chiu RW, Lo YM .(2008) Detection and characterization of placental microRNAs in maternal plasma. Clin Chem 54:482–490.

Cowan ML, Vera J. (2008) Proteomics: Advances in biomarker discovery. Exp Rev Proteomics 5:21–23.

de Winter RJ, Koster RW, Sturk A, Sanders GT. (1995) Value of myoglobin, troponin T, and CK-MBmass in ruling out an acute myocardial infarction in the emergency room. Circulation 92:3401–3407.

Dillhoff M, Wojcik SE, Bloomston M. (2008) MicroRNAs in solid tumors. J Surg Res, in press.

Dorn GW II, Matkovich SJ. (2008) Put your chips on transcriptomics. Circulation 118:216–218.

Dorn GW II, Robbins J, Sugden PH. (2003) Phenotyping hypertrophy: eschew obfuscation. Circ Res 92:1171–1175.

Ebert MP, Korc M, Malfertheiner P, Rocken C. (2006) Advances, challenges, and limitations in serum-proteome-based cancer diagnosis. J Proteome Res 5:19–25.

Gilad S, Meiri E, Yogev Y, Benjamin S, Lebanony D, Yerushalmi N, Benjamin H, Kushnir M, Cholakh H, Melamed N, Bentwich Z, Hod M, Goren Y, Chajut A .(2008) Serum microRNAs are promising novel biomarkers. PLoS ONE 3:e3148.

Hunter MP, Ismail N, Zhang X, Aguda BD, Lee EJ, Yu L, Xiao T, Schafer J, Lee ML, Schmittgen TD, Nana-Sinkam SP, Jarjoura D, Marsh CB. (2008) Detection of microRNA expression in human peripheral blood microvesicles. PLoS ONE 3:e3694.

Jaffe AS, Babuin L, Apple FS. (2006) Biomarkers in acute cardiac disease: the present and the future. J Am Coll Cardiol 48:1–11.

Jaffe AS, Ravkilde J, Roberts R, Naslund U, Apple FS, Galvani M, Katus H. (2000) It's time for a change to a troponin standard. Circulation 102:1216–1220.

Januzzi JL, van KR, Lainchbury J, Bayes-Genis A, Ordonez-Llanos J, Santalo-Bel M, Pinto YM, Richards M. (2006) NT-proBNP testing for diagnosis and short-term prognosis in acute destabilized heart failure: an international pooled analysis of 1256 patients. Eur Heart J 27:330–337.

Ji X, Takahashi R, Hiura Y, Hirokawa G, Fukushima Y, Iwai N. (2009) Plasma miR-208 as a biomarker of myocardial injury. Clin Chem 55:1944–1949.

Laterza OF, Lim L, Garrett-Engele PW, Vlasakova K, Muniappa N, Tanaka WK, Johnson JM, Sina JF, Fare TL, Sistare FD, Glaab WE. (2009) Plasma MicroRNAs as sensitive and specific biomarkers of tissue injury. Clin Chem 55:1977–1983.

Lawrie CH, Gal S, Dunlop HM, Pushkaran B, Liggins AP, Pulford K, Banham AH, Pezzella F, Boultwood J, Wainscoat JS, Hatton CS, Harris AL. (2008) Detection of elevated levels of tumourassociated microRNAs in serum of patients with diffuse large B-cell lymphoma. Br J Haematol 141:672–675.

Lee YS, Dutta A. (2008) MicroRNAs in cancer. Annu Rev Pathol 4:199–227.

Liang Y, Ridzon D, Wong L, Chen C. (2007) Characterization of microRNA expression profiles in normal human tissues. BMC Genomics 8:166.

Lu J, Getz G, Miska EA, Alvarez-Saavedra E, Lamb J, Peck D, Sweet-Cordero A, Ebert BL, Mak RH, Ferrando AA, Downing JR, Jacks T, Horvitz HR, Golub TR. (2005) MicroRNA expression profiles classify human cancers. Nature 435:745-746.

Margulies KB, Bednarik DP, Dries DL. (2009) Genomics, transcriptional profiling, and heart failure. J Am Coll Cardiol 53:1752–1759.

Mitchell PS, Parkin RK, Kroh EM, Fritz BR, Wyman SK, Pogosova-Agadjanyan EL, Peterson A, Noteboom J, O'Briant KC, Allen A, Lin DW, Urban N, Drescher CW, Knudsen BS, Stirewalt DL, Gentleman R, Vessella RL, Nelson PS, Martin DB, Tewari M. (2008) Circulating microRNAs as stable blood-based markers for cancer detection. Proc Natl Acad Sci USA 105:10513–10518.

Persson L, Hardemark HG, Gustafsson J, Rundstrom G, Mendel-Hartvig I, Esscher T, Pahlman S. (1987) S-100 protein and neuron-specific enolase in cerebrospinal fluid and serum: markers of cell damage in human central nervous system. Stroke 18:911–918.

Resnick KE, Alder H, Hagan JP, Richardson DL, Croce CM, Cohn DE. (2009) The detection of differentially expressed microRNAs from the serum of ovarian cancer patients using a novel real-time PCR platform. Gynecol Oncol 112:55–59.

Rothermundt M, Peters M, Prehn JH, Arolt V. (2003) S100B in brain damage and neurodegeneration. Microsc Res Tech 60:614–632.

Taylor DD, Gercel-Taylor C. (2008) MicroRNA signatures of tumor-derived exosomes as diagnostic biomarkers of ovarian cancer. Gynecol Oncol 110:13–21.

Thum T, Galuppo P, Wolf C, Fiedler J, Kneitz S, van Laake LW, Doevendans PA, Mummery CL, Borlak J, Haverich A, Gross C, Engelhardt S, Ertl G, Bauersachs J. (2007) MicroRNAs in the human heart: a clue to fetal gene reprogramming in heart failure. Circulation 116:258–267.

Thygesen K, Alpert JS, White HD. (2007) Universal definition of myocardial infarction. Eur Heart J 28:2525–2538.

Tijsen AJ, Creemers EE, Moerland PD, de Windt LJ, van der Wal AC, Kok WE, Pinto YM. (2010) MiR423-5p As a Circulating Biomarker for Heart Failure. Circ Res 106:1035–1039.

Vaage J, Anderson R (2001) Biochemical markers of neurologic injury in cardiac surgery: the rise and fall of S100beta. J Thorac Cardiovasc Surg 122:853–855.

van Kimmenade RR, Pinto YM, Januzzi JL. (2008) Importance and interpretation of intermediate amino-terminal pro-B-type natriuretic peptide concentrations. Am J Cardiol 101:39–42.

Waldman SA, Terzic A. (2008) MicroRNA signatures as diagnostic and therapeutic targets. Clin Chem 54:943–944.

Wang GK, Zhu JQ, Zhang JT, Li Q, Li Y, He J, Qin YW, Jing Q. (2010) Circulating microRNA: a novel potential biomarker for early diagnosis of acute myocardial infarction in humans. Eur Heart J 31:659–666.

Wang K, Zhang S, Marzolf B, Troisch P, Brightman A, Hu Z, Hood LE, Galas DJ. (2009) Circulating microRNAs, potential biomarkers for drug-induced liver injury. Proc Natl Acad Sci USA 106:4402–4407.

White HD, Chew DP. (2008) Acute myocardial infarction. Lancet 372:570–584.

Xi Y, Formentini A, Chien M, Weir DB, Russo JJ, Ju J, Kornmann M, Ju J. (2006) Prognostic Values of microRNAs in Colorectal Cancer. Biomark Insights 2:113–121.

Yang B, Lin H, Xiao J, Luo X, Li B, Lu Y, Wang H, Wang Z. (2007) The muscle-specific microRNA miR-1 causes cardiac arrhythmias by targeting GJA1 and KCNJ2 genes. Nat Med 13:486–491.

miRNA Interference Approaches for Cardiovascular Disease

Abstract: miRNA interference (miRNAi) is a new concept that I proposed for the development of strategies and technologies to manipulate the function, stability, biogenesis, or expression of miRNAs and as such it indirectly interferes with expression of protein-coding mRNAs. This chapter aims to give an implicitly detailed introduction to some of the miRNAi strategies and technologies developed so far. The fundamental mechanism of miRNA regulation of gene expression is miRNA:mRNA interaction or binding. A key to interfere with miRNA actions is to disrupt the miRNA:mRNA interaction. In order to achieve this aim, one can either manipulate miRNAs or mRNAs to alter the miRNA:mRNA interaction. For miRNAs, one can either mimic miRNA actions to enhance the miRNA:mRNA interaction or to inhibit miRNAs to break the miRNA:mRNA interaction. Both gain-of-function and loss-of-function technologies are necessary tools for understanding miRNAs. The miRNAi technologies have opened up new opportunities for creative and rational designs of a variety of combinations integrating varying nucleotide fragments for various purposes and provided exquisite tools for functional analysis related to identification and characterization of targets of miRNAs and their functions in gene controlling program. In addition, the miRNAi technologies also offer the strategies and tools for designing new agents for gene therapy of human disease.

INTRODUCTION

RNA interference (RNAi) is a well-known strategy for gene silencing; this strategy takes the advantages of the capability of small double-stranded RNA molecules (siRNAs) to bind RNA-induced silencing complex (RISC) on one hand and to bind target genes (mRNAs) on the other hand [Xia *et al.*, 2002; Golden *et al.*, 2008; Pushparaj *et al.*, 2008]. Through such dual interactions, siRNAs elicit powerful knockdown of gene expression by degrading their target mRNAs. Two key characteristics of the RNAi strategy are that the only target of RNAi is mRNAs and that the only outcome of RNAi is silencing of mRNAs. In other words, the RNAi strategy uses siRNAs to interfere directly with mRNAs (mostly protein-coding genes) to silence gene expression.

Taking the same concept, I proposed a new concept of microRNA interference (miRNAi) [Wang *et al.*, 2008; Wang, 2009]. miRNAi manipulates the function, stability, biogenesis, or expression of miRNAs and as such it indirectly interferes with expression of protein-coding mRNAs.

The fundamental mechanism of miRNA regulation of gene expression is miRNA:mRNA interaction or binding. A key to interfere with miRNA actions is to disrupt the miRNA:mRNA interaction. In order to achieve this aim, one can either manipulate miRNAs or mRNAs to alter the miRNA:mRNA interaction. For miRNAs, one can either mimic miRNA actions to enhance the miRNA:mRNA interaction or to inhibit miRNAs to break the miRNA:mRNA interaction (Fig. **1**). Additionally, one can also manipulate mRNA to interrupt the miRNA:mRNA interaction.

Both upregulation and downregulation of miRNAs have been frequently implicated in a variety of pathological conditions. Both gain-of-function and loss-of-function technologies are necessary tools for understanding miRNAs. The miRNAi technologies have opened up new opportunities for creative and rational designs of a variety of combinations integrating varying nucleotide fragments for various purposes and provided exquisite tools for functional analysis related to identification and characterization of targets of miRNAs and their functions in gene controlling program. In addition, the miRNAi technologies also offer the strategies and tools for designing new agents for gene therapy of human disease. These agents will possess a backbone structure in the form of oligoribonucleotides or oligodeoxyribonucleotides. The oligomer fragments generated using miRNAi technologies can be chemically modified to improve stability and constructed into plasmids for easier delivery into organisms to treat diseases. Several innovative miRNAi technologies are introduced below; for detailed information on these technologies, please refer to my recent book <<MicroRNA Interference Technologies>> [Wang, 2009]. These technologies have helped basic research on miRNA targeting and function, and might also help to design and generate new compounds for the treatment of cardiovascular disease associated with miRNAs.

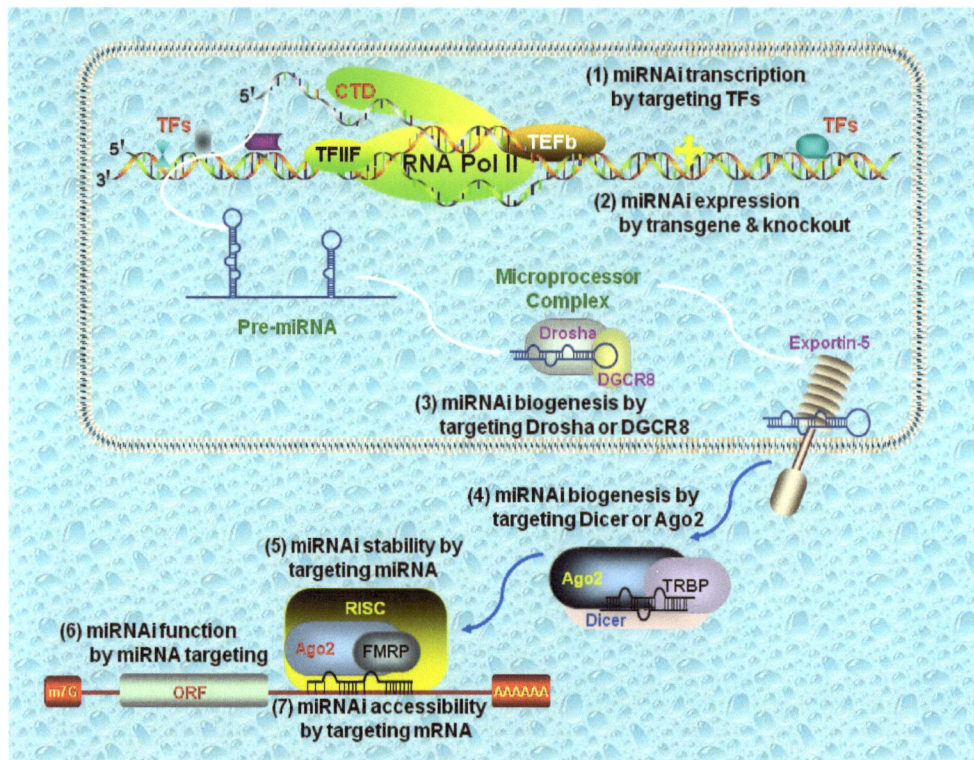

Figure 1: Diagram illustrating the miRNAi strategies: **(1)** miRNAi can be directed to interfere with miRNA transcription by targeting transcriptional factors to either enhance or repress transcription. **(2)** miRNAi can change miRNA expression by creating Transgene or knockout for *in vivo* model studies. **(3)** and **(4)** miRNAi can disrupt miRNA biogenesis by targeting Drosha, DGCR8, Dicer, and Ago2. **(5)** miRNAi can be directed to target the stability of miRNAs to knockdown miRNA level. **(6)** miRNAi can be used to change the function of miRNA by miRNA targeting (miRNA replacement). **(7)** miRNAi can block miRNA accessibility to its binding site in the target genes [Wang, 2009].

SYNTHETIC CANONICAL miRNA TECHNOLOGY

Synthetic Canonical miRNAs are exogenously applied miRNAs in the form of mature double-stranded constructs or of precursor hairpin constructs that are identical in sequence to their counterpart endogenous miRNAs. For convenience, I use the abbreviation SC-miRNAs for Synthetic Canonical miRNAs throughout. SC-miRNAs, once introduced into cells, produce seemingly identical gene regulation and cellular functions as their endogenous miRNA counterparts do. Exogenously transfected miRNAs that can mimic the actions of endogenous canonical miRNAs are the most fundamental and commonly used miRNA-gain-of-function approach.

(1) **To achieve "miRNA-gain-of-function".** Perturbation of miRNA expression, both overexpression and silencing, is a powerful approach to study miRNA functions and to validate miRNA targets. Transient overexpression of miRNAs in cell-based assays can be achieved by transfection and in tissues by *in vivo* gene transfer techniques. This "miRNA-gain-of-function" strategy has become an indispensable and most essential in fundamental research of miRNAs and of miRNA-related biological processes. SC-miRNA is a most direct and straightforward approach to achieve miRNA-gain-of-function. Indeed, ever since the discovery of the first miRNA in 1993 [Lee *et al.*, 1993], SC-miRNAs have been utilized in nearly all experiments involving miRNAs.

(2) **To achieve "miRNA replacement therapy".** Some miRNAs are decreased in their expression levels in cardiac disease (e.g., miR-1 and miR-133 in cardiac hypertrophy). Correction of deregulated expression of these endogenous miRNAs by 'replacement therapy' may be an efficient approach for management of the disorders. Introduction of synthetic miRNAs into cells as a "gain-of-function" approach has proven feasible under many conditions. There has already been a suggestion that replacement of miRNAs may be therapeutically beneficial in some cancers, and other therapeutic opportunities will certainly be uncovered in the future.

miRNA MIMIC TECHNOLOGY

The miRNA Mimic technology (miR-Mimic) is an innovative approach for gene silencing. This approach is to generate non-natural double-stranded miRNA-like RNA fragments. Such a RNA fragment is designed to have its 5'-end bearing a partially complementary motif to the selected sequence in the 3'UTR unique to the target gene. Once introduced into cells, this RNA fragment, mimicking an endogenous miRNA, can bind specifically to its target gene and produce post-transcriptional repression, more specifically translational inhibition, of the gene. Unlike endogenous miRNAs, miR-Mimics act in a gene-specific fashion. The miR-Mimic approach belongs to the "miRNA-targeting" and "miRNA-gain-of-function" strategy, and is primarily used as an exogenous tool to study gene function by targeting mRNA through miRNA-like actions in mammalian cells. This technology was developed by my research group in 2007 [Xiao *et al.*, 2007; Xiao *et al.*, 2008] (Fig. **2**).

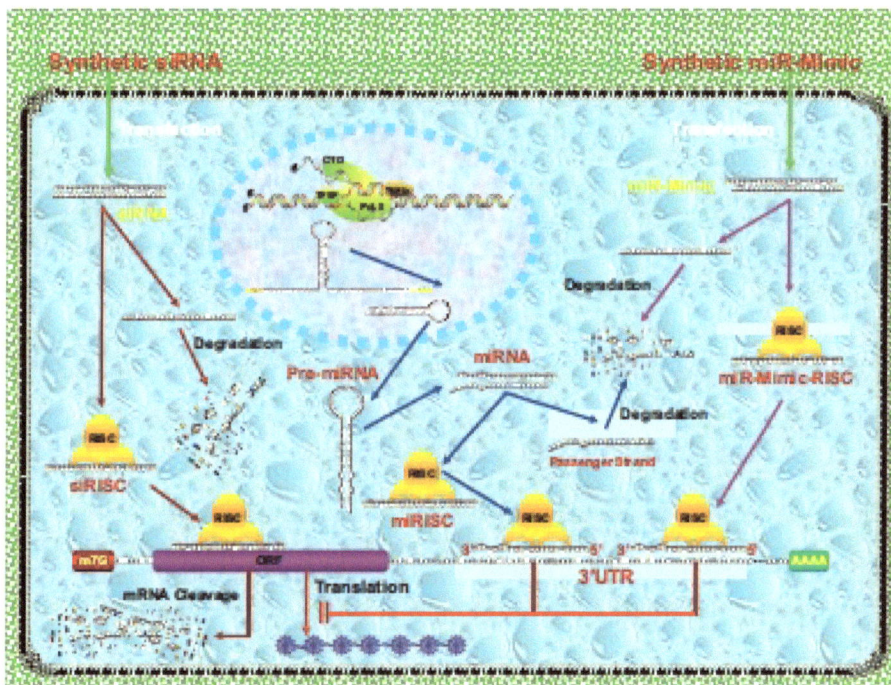

Figure 2: Schematic presentation of actions of miRNA mimic (miR-Mimic) compared with the miRNA and small interference RNA (siRNA). Synthetic miR-Mimic and siRNA are introduced into the cells and endogenous miRNA is synthesized by the cell. siRNAs bind to the coding region of target miRNAs and cause mRNA cleavage; miRNAs bind to 3'UTR of multiple target mRNAs and produce non-gene-specific post-transcriptional repression to inhibit translation; miR-Mimics bind to 3'UTR of unique target mRNAs and produce gene-specific post-transcriptional repression to inhibit translation.

A general and unique feature of the miRNA–target RNA interaction is imperfect complementarity between the miRNA guide strand and its target mRNA. Hence, miRNA guide strands usually form bulge structures due to mismatches with its target sequence. The sequence specificity for target recognition by the guide miRNA strand is determined by nucleotides 2–8 of its 5' region, referred to as the "seed site" [Doench & Sharp, 2004; Lewis *et al.*, 2003]. This short seed site required for miRISC function raises the potential for a single miRNA to target multiple mRNAs. Indeed, it has been confirmed that unlike a non-natural siRNA which targets a particular gene, each single endogenous miRNA has the potential of regulating multiple protein-coding genes, as many as 1000. On the other hand, each individual gene may be regulated by multiple miRNAs. This implies that the actions of miRNAs are not gene-specific, but sequence-specific for they can act on any genes carrying motifs matching their seed sites. Thus, when aiming to silence a particular gene using a naturally occurring miRNA, one may actually knockdown a group of genes. This property of miRNAs creates a hurdle for exploiting miRNA function and targets. To this end, we have developed a novel technology called microRNAs Mimics or miR-Mimics technology. The miR-Mimics generated using this approach act by miRNA mechanisms in a gene-specific manner.

(1) **To achieve gene silencing by miRNA mechanism in a gene-specific manner.** As already mentioned in the previous chapter, miRNA action is not gene-specific. Thus, the gene-silencing action through SC-miRNA strategy is not gene-specific either. The miR-Mimic technology was developed to circumvent this limitation. This property of miR-Mimics can be advantageous over SC-miRNAs when specific-genes need to be knocked down which happens in many situations.

(2) **To complement the loss of endogenous miRNAs under certain conditions.** In some abnormal conditions, some miRNAs are downregulated in their expression, leading to aberrantly enhanced expression of some protein-coding genes causing diseased phenotypes. Replacement of these miRNAs may reverse this process. Alternatively, application of miR-Mimics targeting the disease-causing genes to prevent their upregulation may be an efficient maneuver to tackle the problem. For example, in the development of cancer, expression of oncogenes is overexpression, partially as a result of the downregulation of their regulating miRNAs. Under such a situation, use of miR-Mimics to keep down of the oncogenes may help to slow the tumorigenesis.

(3) **Examples of applications.** We have examined the application of the technique to cardiac pacemaker genes HCN2 and HCN4 [Xiao *et al.*, 2007]. Following the protocols described above, we first identified a stretch of sequence in the 3'UTR unique to the HCN2 (or HCN4) gene that is expectedly long enough for miRNA action. Based on the unique sequence we designed a 22-nt miR-Mimic that at the 5' end has eight nucleotides (nucleotides 2–8), and at the 3'-end has seven nucleotides, complementary to the HCN2 (or HCN4) sequence. These miR-Mimics produced substantial repression of HCN channel protein expression with concomitant depression of pacemaker activities and reduction of beating rate of cultured neonatal myocytes but with minimal effects on their mRNA levels. These cardiac automaticity-targeting miR-Mimics are expected to act like heart rate-reducing agents that have been shown to be beneficial to cardiac function during infarction and to be able to suppress ectopic beats that can be elicited by abnormal automaticity. The results demonstrated a promise of utilizing the technology for gene-specific repression of expression at the protein level based on the principle of miRNA actions.

ANTI-miRNA ANTISENSE OLIGONUCLEOTIDES TECHNOLOGY

One of the indispensable approaches in miRNA research as well as in miRNA therapy is inhibition or loss-of-function of miRNAs. Multiple steps in pathway for miRNA biogenesis could be targeted for inhibition of miRNA production and maturation. Thus far, nonetheless, the anti-miRNA antisense inhibitor oligoribonucleotides (AMO) technology used to target mature miRNAs has found most value for these applications. Standard AMO is a single-stranded 2'-O-methyl (2'-OMe)-modified oligoribonucleotide fragment exactly antisense to its target miRNA. The AMO technology was initially established in 2004 by Tuschl's laboratory [Meister *et al.*, 2004] and by Zamore's laboratory [Hutvágner *et al.*, 2004]. Since then, AMO technology has undergone many important modifications to enhance the efficiency and specificity of miRNA interference. These include a cholesterol moiety-conjugated 2'-OMe modified AMOs called antagomiR [Krutzfeldt *et al.*, 2005], locked nucleic acid (LNA)-modified AMOs [Ørom *et al.*, 2006; Davis *et al.*, 2006], 2'-O-methoxyethyl (2'-MOE) [Esau *et al.*, 2006; Davis *et al.*, 2006], 2'-flouro (2'-F) [Davis *et al.*, 2006], phosphorothioate backbone modification and a peptide nucleic acid (PNA)-modified AMOs [Fabani & Gait, 2008]. For the sake of clarity and convenience, I here designate all different types of anti-miRNA antisense AMO and the conventional anti-mRNA antisense ASO, though ASO has also been used to refer to anti-miRNA antisense by some authors.

Multiple steps along the path of biogenesis of miRNAs can be interfered to achieve miRNA knockdown (Fig. **1**). The base-pair interaction between miRNAs and mRNAs is essential for the function of miRNAs; therefore, a logical approach of silencing miRNAs is to use a nucleic acid that is antisense to the miRNA. These anti-miRNA oligonucleotides (AMOs) specifically and stoichiometrically bind, and efficiently and irreversibly silence, their target miRNAs. It is the most straightforward and apparently most effective strategy tested so far to block the function of miRNAs in RISC. It mediates potent and miRNA-specific inhibition of miRNA function, providing a powerful "loss-of-function" strategy for interfering miRNA expression.

Among the various forms of AMOs, antagomiR that has its end conjugated with a cholesterol moiety has demonstrated most impressive effectiveness against target miRNAs, intracellular stability, particularly for *in vivo* applications [Krutzfeldt *et al.*, 2005]. This particular form of AMO will be discussed more in detail in a later section of this chapter.

(1) **To validate the miRNA targets.** Understanding the biological function of miRNAs requires knowledge of their mRNA targets. To validate the theoretically predicted miRNA targets, new and rapid methods for sequence-specific inactivation of miRNAs are needed. The ability of an AMO to block the binding of its targeted miRNA to a gene helps determining whether this gene is a target of that miRNA. The method thus provides a means of connecting miRNA target identification with biological function and of distinguishing phenocritical from neutral targets.

(2) **To validate miRNA function.** The AMO technique can be used not only for genome-wide phenotypic screening, but also for a detailed subsequent characterization of phenotypes using markers and epistasis experiments, thereby placing miRNAs more precisely within biological processes and pathways. Use of AMOs can aid to define the cause-effect relationship between a miRNA and a biological event, thereby the cellular function of miRNAs.

(3) **To achieve upregulation of the cognate target protein.** The AMO technology is unique in its action and outcome of action. AMOs act on miRNAs in a 'loss-of-function' manner to silence these miRNAs, but they result in 'gain-of-function' of target protein-coding genes via relieving the post-transcriptional repressive effects of their targeted miRNAs on these protein-coding genes. In this way, AMOs up-regulate gene expression.

(4) This property of AMOs can be utilized in many situations for the benefits of treating human disease in cases when up-regulation of gene expression is desirable. One most obvious application of the AMO approach is to up-regulate expression of cytoprotective molecules that are repressed by miRNAs in myocardial ischemia.

(5) 'Gain-of-function' of target protein-coding genes produced by AMOs can also be used to validate the cellular function and target gene of miRNAs. If an AMO up-regulates tumor suppressor gene, then it is very likely that the counterpart of this AMO, the targeted miRNA, acts to repress the expression of this tumor suppressor gene.

(6) **To reverse the pathological process.** Many miRNAs have been implicated in human disease or in animal models of disease, such as cancer, cardiovascular disturbances (e.g. ischemic arrhythmogenesis, cardiac hypertrophy, and heart failure), metabolic disorders (e.g. type 2 diabetes), viral diseases, Alzheimer's disease, etc. Targeting pertinent miRNAs with AMOs has been shown to be effective in reversing these pathological processes. For example, we have demonstrated the ability of AMOs to antagonize the ischemic arrhythmias induced by miR-1 [Yang *et al.*, 2007], to abolish the abnormal QT prolongation caused by miR-133 in diabetic hearts [Xiao *et al.*, 2007a], and to reverse miR-1-induced heart rate-reducing effects [Xiao *et al.*, 2007b].

MULTIPLE-TARGET ANTI-miRNA ANTISENSE OLIGONUCLEOTIDES TECHNOLOGY

Anti-miRNA antisense inhibitors (AMOs) have demonstrated their utility in miRNA research and potential in miRNA therapy. However, it has become clear that a particular condition may be associated with multiple miRNAs and a given gene may be regulated by multiple miRNAs. For example, a study directed to the human heart identified 67 significantly up-regulated miRNAs and 43 significantly down-regulated miRNAs in failing left ventricles versus normal hearts [Thum *et al.*, 2007]. No less than 5 different miRNAs have been shown to critically involve in cardiac hypertrophy. Similarly, Volinia *et al* [2006] conducted a large-scale miRNA analysis on 540 solid tumor samples including lung, breast, stomach, prostate, colon, and pancreatic tumors. Their survey revealed 15 miRNAs up-regulated and 12 down regulated in human breast cancer tissues. Similar changes of multiple miRNAs were also found in other five solid tumor types. Many of these miRNAs have been reported to regulate cell proliferation or apoptosis and some of them have been considered oncogenic miRNAs or tumor suppressor miRNAs. These facts raised two questions. If targeting a single miRNA is sufficient for tackling a particular pathological condition? If simultaneously targeting multiple miRNAs relevant to a particular condition offers an improved outcome than targeting a single miRNA using the regular AMO techniques?

These properties of miRNA regulation may well create some uncertainties of outcomes by applying the AMO technology to silence miRNAs since knocking down a single miRNA may not be sufficient to achieve the expected interference of cellular process and gene expression which are regulated by multiple miRNAs. Tools that allow multi-miRNA knockdown will be essential for identification and validation of miRNA targets. For such applications, co-transfection of multiple AMOs targeting various isoforms is possible [Bommer *et al.*, 2007]. In this respect, genetic approaches are superior for studying individual miRNA family members, whereas miRNA sponges (to be introduced in the next chapter) or multiple AMOs are appropriate for studying miRNA families whose members only contain a common seed sequence; single AMO studies may simplify the study of nearly identical miRNA paralogs. Combinations of AMOs targeting unrelated miRNAs have also been used to disrupt more than one miRNA in the same transfected cells, obviating the need to make and combine multiple genetic knockouts. Because combinatorial control of targets by miRNAs may be common [Bartel & Chen, 2004], this approach may prove particularly important for uncovering networks of miRNAs that act together. Vermeulen *et al* [2007] showed that co-transfection of a six-AMO mixture can effectively derepress reporters for each individual miRNA. Functional studies also suggest that co-transfection of AMOs is effective. For instance, Pedersen *et al* [2007] tested the efficacy of five interferon-β induced miRNAs with seed matches to Hepatitis C genes in antiviral activity; indeed, simultaneous co-transfection of all five AMOs, but not controls, significantly enhanced Hepatitis C RNA production.

Co-application of multiple AMOs while effective in some cases may be problematic in that control of equal transfection efficiency is difficult, if not impossible. To tackle the problem, an innovative strategy, the multiple-target AMO technology or MT-AMO technology, which confers a single AMO fragment the capability of targeting multiple miRNAs has been developed in my laboratory [Lu *et al.*, 2009]. This modified AMO carries multiple antisense units which are engineered into a single unit that is able to simultaneously silence multiple target miRNAs. Studies suggest the MT-AMO an improved approach for miRNA target-gene finding and for studying function of miRNAs. This novel strategy may find its broad application as a useful tool in miRNA research for exploring biological processes involving multiple miRNAs and multiple genes, and the potential as a miRNA therapy for human disease such as cancer and cardiac disorders.

miRNA SPONGE TECHNOLOGY

miRNA Sponge technology is an innovative approach used to generate RNAs containing multiple, tandem binding sites for a miRNA seed family of interest and able to target *all* members of that miRNA seed family. When vectors encoding the miRNA sponges are transiently transfected into cultured cells, they depress miRNA targets as strongly as the conventional AMOs described in Chapter 6. The major advancement of this technique over the AMO technique is that it can better inhibit functional classes of miRNAs than do AMOs that are designed to block single miRNA sequences. The main principle of the miRNA Sponge technology is identical to that the MT-AMO technology described in Chapter 7: targeting multiple miRNAs. The miRNA Sponge technology was established by Sharp's laboratory in 2007 [Ebert *et al.*, 2007; Hammond, 2007]. Similar to the AMO approach, miRNA Sponge technology belongs to the "targeting-miRNA" and "miRNA-loss-of-function" strategy. The miRNA Sponge technology complies with the 'Single-Drugn, Multiple-Target' concept [Gao *et al.*, 2006].

Many miRNAs are members of families that share a seed sequence, but may have one or more nucleotide changes in the remaining sequence [Bommer *et al.*, 2007; Pedersen *et al.*, 2007; Vermeulen *et al.*, 2007]. Moreover, many miRNAs are expressed from multiple genomic loci. To achieve adequate miRNA-loss-of-function for elucidating a certain cellular process, the conventional AMO strategy falls short in dealing with multiple miRNAs. On the other hand, creating genetic knockouts to determine the function of miRNA families is difficult, as individual miRNAs expressed from multiple genomic loci or from multiple members of a same miRNA seed family may repress a common set of targets containing a complementary seed sequence. Thus, a method for inhibiting these functional classes of paralogous miRNAs *in vivo* is needed.

For this reason, Ebert *et al* [2007] invented an innovative anti-miRNA approach termed 'miRNA sponges'. The idea behind it is to produce a single specie of RNAs containing multiple, tandem binding sites for a miRNA seed family of interest, in order to target *all* members of that miRNA seed family, taking advantage of the fact that the interaction between miRNA and target is nucleated by and largely dependent on base-pairing in the seed region (positions 2–8 of the miRNA). The authors constructed sponges by inserting tandemly arrayed miRNA binding sites into the 3'UTR of a reporter gene encoding destabilized GFP driven by the CMV promoter, which can yield abundant expression of the competitive inhibitor transcripts. Like the AMO and the MT-AMO approaches, the miRNA Sponge technology is able to interfere with the function of natural, endogenous target miRNAs thereby the target genes of the miRNAs, and can be used for target validation and phenotypic analysis.

miRNA-MASKING ANTISENSE OLIGONUCLEOTIDES TECHNOLOGY

miRNA-Masking Antisense Oligonucleotides Technology (miR-Mask) is an AMO approach of different sort. A standard miR-Mask is single-stranded 2'-*O*-methyl-modified oligoribonucleotide (or other chemically modified) that is a 22-nt antisense to a protein-coding mRNA as a target for an endogenous miRNA of interest. Instead of binding to the target miRNA like an AMO, a miR-Mask does not directly interact with its target miRNA but binds to the binding site of that miRNA in the 3'UTR of the target mRNA by fully complementary mechanism. In this way, the miR-Mask covers up the access of its target miRNA to the binding site so as to derepress its target gene (mRNA) via blocking the action of its target miRNA. The anti-miRNA action of a miR-Mask is gene-specific because it is designed to be fully complementary to the target mRNA sequence of a miRNA. The anti-miRNA action of a miR-Mask is also miRNA-specific as well because it is designed to target the binding site of that particular miRNA. The miR-Mask approach is a valuable supplement to the AMO technique; while AMO is indispensable for studying the overall function of a miRNA, the miR-Mask might be more appropriate for studying the specific outcome of regulation of the target gene by the miRNA. This technology was first established by my research group in 2007 [Xiao *et al.*, 2007b] and a similar approach with the same concept was subsequently reported by Schier's laboratory [Choi *et al.*, 2007]. Similar to the AMO approach, miR-Mask technology belongs to the "targeting-miRNA" and "miRNA-loss-of-function" strategy.

Each single miRNA may regulate as many as 1000 protein-coding genes and each gene may be regulated by multiple miRNAs. This implies that the action of miRNAs is binding sequence-specific but not gene-specific; similarly, the action of AMO, thereby MT-AMO and miRNA Sponge, is miRNA-specific but not gene-specific either. These properties of miRNAs and AMOs may present the obstacles for development as therapeutic agents since they may elicit unwanted side effects and toxicity through their non-gene-specific functional profiles. For example, the muscle-specific miRNA miR-1 has the potential to post-transcriptionally repress a number of ion channel genes including cardiac sodium channel gene SCNCA5, pacemaker channel gene HCN4, gap junction channel connexin 43, inward rectifier K^+ channel KCNJ2, voltage-dependent K^+ channel KCND2. Based on this targeting, miR-1 is expected to affect cardiac electrophysiology. Taking the concept of "miRNA as a Regulator of a Cellular Function", one can just focus on what miR-1 does on the cardiac electrophysiology. However, if one wants to understand the mechanisms using miR-1 loss-of-function strategy, the AMO, MT-AMO, and miR-Sponge technologies will all fall short due to their lack of gene specificity. By knocking down miR-1, one will potentially alter the expression of all miR-1 target genes mentioned above.

To emasculate the problem, we have developed the miRNA-Masking Antisense Oligonucleotides Technology (miR-Mask) that provides a gene-specific strategy for studying miRNA function and mechanisms. Using the miR-Mask approach, one is now able to dissect the role of each of the target genes of a miRNA, say the ion channel genes for miR-1. For instance, one can use a miR-Mask on SCNCA5 to explore the role of miR-1 regulation of this sodium channel on cardiac electrophysiology. Soon after our publication on this technology, Choi *et al* [2007] also published a study using essentially the same strategy, and they named the technology "Target Protector".

The miR-Mask technology is an alternative to the AMO approach. But unlike AMO that acts in a non-gene-specific manner, the miR-Mask finds its particular value in targeting miRNA in a gene-specific fashion (Fig. **3**). It is particularly useful when inhibiting miRNA action on a particular protein-coding gene without affecting the level of this miRNA and its silencing effects on other genes is required.

We have validated the miR-Mask technology by testing its application to the cardiac pacemaker channel-encoding genes HCN2 and HCN4 [Xiao *et al.*, 2007]. We created miR-Masks that are able to bind to HCN2 and HCN4 and prevent the repressive actions of miR-1 and miR-133. These miR-Masks resulted in enhanced protein expression of the pacemaker channels and increased pacemaker activities revealed by whole-cell patch-clamp recordings.

Functionally, the miR-Masks for HCN channels cause acceleration of heart rate in rats, simulating "biological pacemakers" [Xiao *et al.*, 2007].

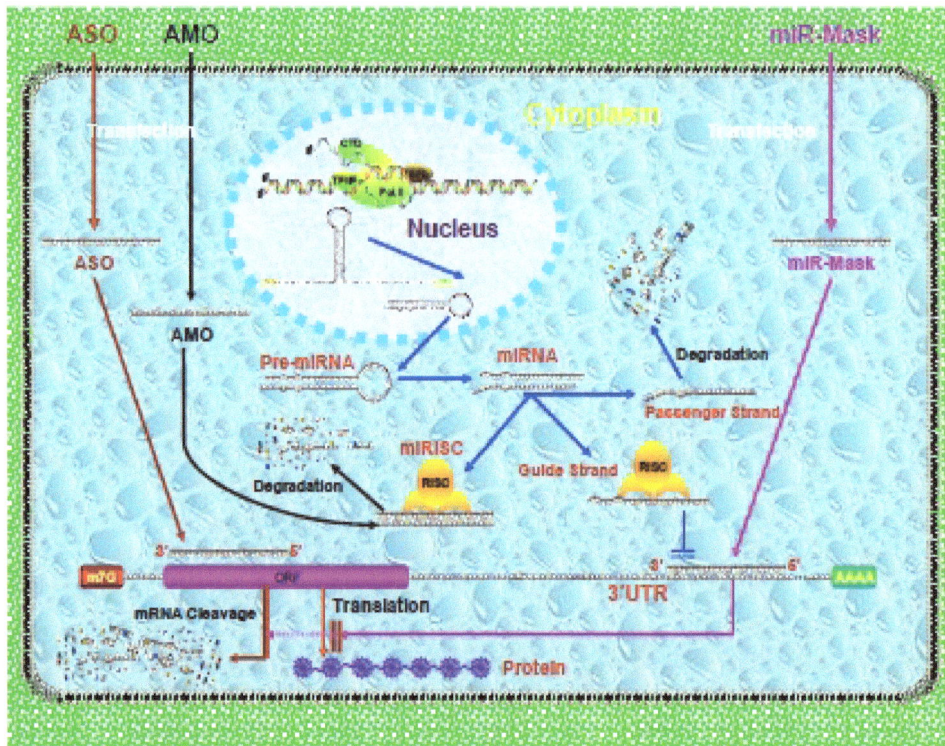

Figure 3: Schematic presentation of actions of miRNA-masking antisense oligonucleotide (miR-Mask) compared with the conventional antisense oligodeoxynucleotide (ASO) and anti-miRNA antisense inhibitor oligonucleotide (AMO) technologies. Synthetic nucleic acids are introduced into the cells. ASO bind to the coding region of the target mRNA and hinder the translation process; AMOs bind to the target miRNA, resulting in miRNA cleavage; miR-Masks bind to the binding site of miRNAs in 3'UTR of the target mRNA and prevent miRNAs from binding to the target mRNA, leading to a relief of translational repression without affecting miRNAs.

REFERENCES

Bartel DP, Chen CZ. (2004) Micromanagers of gene expression: the potentially widespread influence of metazoan microRNAs. Nat Rev Genet 5:396–400.

Bommer GT, Gerin I, Feng Y, Kaczorowski AJ, Kuick R, Love RE, Zhai Y, Giordano TJ, Qin ZS, Moore BB, MacDougald OA, Cho KR, Fearon ER. (2007) p53-mediated activation of miRNA34 candidate tumor-suppressor genes. Curr Biol 17:1298–1307.

Choi WY, Giraldez AJ, Schier AF. (2007) Target protectors reveal dampening and balancing of Nodal agonist and antagonist by miR-430. Science 318:271–274.

Davis S, Lollo B, Freier S, Esau C. (2006) Improved targeting of miRNA with antisense oligonucleotides. Nucleic Acids Res 34:2294–2304.

Doench JG, Sharp PA. (2004) Specificity of microRNA target selection in translational repression. Genes Dev 18:504–511.

Ebert MS, Neilson JR, Sharp PA. (2007) MicroRNA sponges: competitive inhibitors of small RNAs in mammalian cells. Nat Methods 4:721–726.

Esau C, Davis S, Murray SF, Yu XX, Pandey SK, Pear M, Watts L, Booten SL, Graham M, McKay R, Subramaniam A, Propp S, Lollo BA, Freier S, Bennett CF, Bhanot S, Monia BP. (2006) miR-122 regulation of lipid metabolism revealed by *in vivo* antisense targeting. Cell Metab 3:87–98.

Fabani MM, Gait MJ. (2008) miR-122 targeting with LNA/2'-O methyl oligonucleotide mixmers, peptide nucleic acids (PNA), and PNA-peptide conjugates. RNA 14:336–346.

Gao H, Xiao J, Sun Q, Lin H, Bai Y, Yang L, Yang B, Wang H, Wang Z. (2006) A single decoy oligodeoxynucleotides targeting multiple oncoproteins produces strong anticancer effects. Mol Pharmacol 70:1621–1629.

Golden DE, Gerbasi VR, Sontheimer EJ. (2008) An inside job for siRNAs. Mol Cell 31:309–312.

Hammond SM. (2007) Soaking up small RNAs. Nat Methods 4:694–695.

Hutvágner G, Simard MJ, Mello CC, Zam PD. (2004) Sequence-Specific Inhibition of Small RNA Function. PLoS Biol 2:465–475.

Krutzfeldt J, Rajewsky N, Braich R, Rajeev KG, Tuschl T, Manoharan M, Stoffel M. (2005) Silencing of microRNAs *in vivo* with 'antagomirs'. Nature 438:685–689.

Meister G, Landthaler M, Dorsett Y, Tuschl T. (2004) Sequence-specific inhibition of microRNA- and siRNA-induced RNA silencing. RNA 10:544–550.

Lee RC, Feinbaum RL, Ambros V. (1993) The *C. elegans* heterochronic gene lin-4 encodes small RNAs with antisense complementarity to lin-14. Cell 75:843–854.

Lewis BP, Shih IH, Jones-Rhoades MW, Bartel DP, Burge CB. (2003) Prediction of mammalian microRNA targets. Cell 115:787–798.

Lu Y, Xiao J, Lin H, Bai Y, Luo X, Wang Z, Yang B. (2009) Complex antisense inhibitors offer a superior approach for microRNA research and therapy. Nucleic Acids Res 37:e24–e33.

Ørom UA, Kauppinen S, Lund AH. (2006) LNA-modified oligonucleotides mediate specific inhibition of microRNA function. Gene 372:137–141.

Pedersen IM, Cheng G, Wieland S, Volinia S, Croce CM, Chisari FV, David M. (2007) Interferon modulation of cellular microRNAs as an antiviral mechanism. Nature 449:919–922.

Pushparaj PN, Aarthi JJ, Manikandan J, Kumar SD. (2008) siRNA, miRNA, and shRNA: *in vivo* applications. J Dent Res 87:992–1003.

Thum T, Galuppo P, Wolf C, Fiedler J, Kneitz S, van Laake LW, Doevendans PA, Mummery CL, Borlak J, Haverich A, Gross C, Engelhardt S, Ertl G, Bauersachs J. (2007) MicroRNAs in the human heart: a clue to fetal gene reprogramming in heart failure. Circulation 116:258–267.

Vermeulen A, Robertson B, Dalby AB, Marshall WS, Karpilow J, Leake D, Khvorova A, Baskerville S. (2007) Double-stranded regions are essential design components of potent inhibitors of RISC function. RNA 13:723–730.

Volinia S, Calin GA, Liu CG, Ambs S, Cimmino A, Petrocca F, Visone R, Iorio M, Roldo C, Ferracin M, Prueitt RL, Yanaihara N, Lanza G, Scarpa A, Vecchione A, Negrini M, Harris CC, Croce CM. (2006) A microRNA expression signature of human solid tumors defines cancer gene targets. Proc Natl Acad Sci USA 103:2257–2261.

Wang Z. (2009) MicroRNA Interference Technologies. Springer-Verlag, New York, USA.

Wang Z, Luo X, Lu Y, Yang B. (2008) miRNAs at the heart of the matter. J Mol Med 86:771–783.

Yang B, Lin H, Xiao J, Luo X, Li B, Lu Y, Wang H, Wang Z. (2007) The muscle-specific microRNA miR-1 causes cardiac arrhythmias by targeting GJA1 and KCNJ2 genes. Nat Med 13:486–491.

Xia H, Mao Q, Paulson HL, Davidson BL. (2002) siRNA-mediated gene silencing *in vitro* and in vivo. Nat Biotechnol 20:1006–1010.

Xiao J, Luo X, Lin H, Xu C, Gao H, Wang H, Yang B, Wang Z. (2007a) MicroRNA miR-133 represses HERG K$^+$ channel expression contributing to QT prolongation in diabetic hearts. J Biol Chem 282:12363–12367.

Xiao J, Yang B, Lin H, Lu Y, Luo X, Wang Z. (2007b) Novel approaches for gene-specific interference via manipulating actions of microRNAs: examination on the pacemaker channel genes *HCN2* and *HCN4*. J Cell Physiol 212:285–292.

Index